T-13
4

Chemical Biotechnology and Bioengineering

RSC Green Chemistry

Editor-in-Chief:
Professor James Clark, *Department of Chemistry, University of York, UK*

Series Editors:
Professor George A. Kraus, *Department of Chemistry, Iowa State University, Ames, Iowa, USA*
Professor Andrzej Stankiewicz, *Delft University of Technology, The Netherlands*
Professor Peter Siedl, *Federal University of Rio de Janeiro, Brazil*
Professor Yuan Kou, *Peking University, China*

How to obtain future titles on publication:
A standing order plan is available for this series. A standing order will bring
delivery of each new volume immediately on publication.

For further information please contact:
Book Sales Department, Royal Society of Chemistry, Thomas Graham House,
Science Park, Milton Road, Cambridge, CB4 0WF, UK
Telephone: +44 (0)1223 420066, Fax: +44 (0)1223 420247
Email: booksales@rsc.org
Visit our website at www.rsc.org/books

Chemical Biotechnology and Bioengineering

Xuhong Qian
East China University of Science & Technology, Shanghai, China
Email: xhqian@ecust.edu.cn

Zhenjiang Zhao
East China University of Science & Technology, Shanghai, China
Email: zhjzhao@ecust.edu.cn

Yufang Xu
East China University of Science & Technology, Shanghai, China
Email: yfxu@ecust.edu.cn

Jianhe Xu
East China University of Science & Technology, Shanghai, China
Email: jianhexu@ecust.edu.cn

Y.-H. Percival Zhang
Cell Free Bioinnovations Inc., Blacksburg VA, USA
Email: ypzhang@vt.edu

Jingyan Zhang
East China University of Science & Technology, Shanghai, China
Email: jyzhang@ecust.edu.cn

Yangchun Yong
Jiangsu University, Zhenjiang, China
Email: ycyong@ujs.edu.cn

Fengxian Hu
East China University of Science & Technology, Shanghai, China
Email: hufx@ecust.edu.cn

THE QUEEN'S AWARDS
FOR ENTERPRISE:
INTERNATIONAL TRADE
2013

RSC Green Chemistry No. 34

Print ISBN: 978-1-84973-810-1
PDF eISBN: 978-1-78262-012-9
ISSN: 1757-7039

A catalogue record for this book is available from the British Library

Published by The Royal Society of Chemistry,
Thomas Graham House, Science Park, Milton Road,
Cambridge CB4 0WF, UK

Registered Charity Number 207890

For further information see our web site at www.rsc.org

Preface

The rapid development of modern science and technology is, in many cases, a result of highly interdisciplinary integration. For instance, chemical biology, an interdiscipline of traditional chemistry, biology, and physics, involves the application of chemical techniques, tools, and analyses, and often compounds produced through synthetic chemistry, to the study and manipulation of biological systems. The rapid development of chemical biology prompts more frontier disciplines, such as synthetic biology, chemical biotechnology, and chemical engineering. In particular, with the growing global population as well as energy and environmental crises, using the combination of different disciplines makes it possible to solve some problems that are otherwise hard to resolve by one specific discipline.

Chemical biotechnology and bioengineering utilize small chemical molecules to manipulate bioprocesses. The modified processes will make it possible to generate some chemicals more ecologically and economically, overcoming the resources and environmental crises that traditional chemical biology faces. It could also strengthen the natural bioprocesses that are not presently robust for practical application. This book attempts to show some examples of chemical biotechnology and engineering from macro-biomolecules to cells and whole plants, including chemical modulators in enzymatic reactions; chemical regulators in the regulation of non-canonical DNA structures; chemical biomimetic cofactors *in vitro* biosystems in the production of high-value chemicals and low-value bio-commodities; some chemicals in microbial electrochemical systems for the improvement of the performance/efficiency of extracellular electron transfer between the bacteria and the electrode; elicitors in plant cell culture for precious natural products; and plant activators in crop protection. Attention is paid not only to the chemicals that

RSC Green Chemistry No. 34
Chemical Biotechnology and Bioengineering
By Xuhong Qian, Zhenjiang Zhao, Yufang Xu, Jianhe Xu, Y.-H. Percival Zhang, Jingyan Zhang, Yangchun Yong, and Fengxian Hu
© X.-H. Qian, Z.-J. Zhao, Y.-F. Xu, J.-H. Xu, Y.-H. P. Zhang, J.-Y. Zhang, Y.-C. Yong and F.-X. Hu, 2015
Published by the Royal Society of Chemistry, www.rsc.org

can affect the biological processes, but also to solving some major issues, by introducing some application examples in this book.

To date, few books have been published in these areas; this publication will fill a gap and also catch the hot topic at the interface of chemistry and biotechnology or bioengineering. Consequently, the book should appeal to members of the chemistry as well as the biology and medical communities.

Finally, I thank all the chapter contributors. Without their hard work, the appearance of this book might never have occurred. I, on behalf of all contributors to this book, must express my heartfelt gratitude to the chance that the Royal Society of Chemistry has given us to introduce some of our work. We are also indebted to the editorial team for patiently keeping us on time and suggesting some revisions to our book.

Xuhong Qian

Contents

RSC Green Chemistry No. 34
Chemical Biotechnology and Bioengineering
By Xuhong Qian, Zhenjiang Zhao, Yufang Xu, Jianhe Xu, Y.-H. Percival Zhang, Jingyan Zhang,
Yangchun Yong, and Fengxian Hu
© X.-H. Qian, Z.-J. Zhao, Y.-F. Xu, J.-H. Xu, Y.-H. P. Zhang, J.-Y. Zhang, Y.-C. Yong and F.-X. Hu, 2015
Published by the Royal Society of Chemistry, www.rsc.org

CHAPTER 1

From Chemical Biology to Its Technology and Engineering

XUHONG QIAN*[a], ZHENJIANG ZHAO[b], JIAN-HE XU[c],
YUFANG XU[a,b], YI-HENG PERCIVAL ZHANG[d], FENGXIAN HU[e],
JINGYAN ZHANG[a], AND YANG-CHUN YONG[f]

[a]Shanghai Key Laboratory of Chemical Biology and State Key Laboratory of
Bioreactor Engineering, School of Pharmacy, East China University of Science
and Technology, Shanghai 200237, China; [b]Shanghai Key Laboratory of New
Drug Design, School of Pharmacy, East China University of Science and
Technology, Shanghai 200237, China; [c]Laboratory of Biocatalysis and
Synthetic Biotechnology, State Key Laboratory of Bioreactor Engineering,
East China University of Science and Technology, Shanghai 200237, China;
[d]Cell Free Bioinnovations Inc., 2200 Kraft Drive, Suite 1200B, Blacksburg,
VA 24060, USA; [e]School of Biotechnology, State Key Laboratory of Bioreactor
Engineering, East China University of Science and Technology, Shanghai
200237, China; [f]Biofuels Institute, School of the Environment, Jiangsu
University, Zhenjiang 212013, Jiangsu Province, China
*E-mail: xhqian@ecust.edu.cn

1.1 The History and Applications of Biotechnology

The term 'biotechnology' first appeared in Manchester University, England,
early in the 20th century.[1] In 1912, C. Weizmann isolated a strain of *Clostridium acetobutylicum* that converted carbohydrate into butanol, acetone,
and ethanol.[2] In 1923, T. Walker commenced undergraduate education in

RSC Green Chemistry No. 34
Chemical Biotechnology and Bioengineering
By Xuhong Qian, Zhenjiang Zhao, Yufang Xu, Jianhe Xu, Y.-H. Percival Zhang, Jingyan Zhang,
Yangchun Yong, and Fengxian Hu

the department of fermentation industries; the name of this department was then changed to industrial biochemistry, similar to 'biotechnology'.[1] K. Ereky, a Hungarian engineer and the founding father of biotechnology,[3] first coined the word "biotechnology" in a book published in Berlin in 1919 called *Biotechnologie der Fleisch-, Fett- und Milcherzeugung im landwirtschaftlichen Grossbetriebe* (Biotechnology of Meat, Fat and Milk Production in an Agricultural Large-Scale Farm), in which he explained a bioprocess technology that converts raw materials into a more valuable product. He further developed this concept for the 20th century: Biotechnology means solutions to many social and natural crises, such as food and energy shortages.[2]

In fact, biotechnology as a methodology has existed for a long time, although the term "biotechnology" did not exist then. Around 8000 BC yeast was used to make wheat wine by the Chinese, and also around 6000 BC by the Sumerians and Babylonians.[4] Around 4000 BC the Egyptians used yeast to bake leavened bread. The ancient Chinese produced copious amounts of liquor, rice wine, soy sauce, and vinegar with the earliest biotechnology utilizing conversion by some microbes.[5]

In AD 1673 A. v. Leeuwenhoek, a Dutch scientist known as "the father of microbiology", pointed out the functions of microorganisms in fermentation[6] and made contributions towards handcrafted microscopes and the establishment of microbiology. He was the first one to observe and describe single-celled organisms as *animalcules*, which are now referred to as microorganisms.

Vaccines, which protect people from some terrible diseases, are a typical biotechnology.[7] In the 16th century, the Chinese inoculated smallpox scab powder on the body to protect from smallpox infection. E. Jenner (1749–1823), an English physician known as "the father of immunology", heard a dairymaid's story and started to study cowpox vaccine. L. Pasteur (1822–1895), the other founder of medical microbiology, was famous for his creation of two vaccines for rabies and anthrax.

A. Fleming (1881–1955), a Scottish scientist, discovered the antibiotic penicillin, a secondary metabolite from *Penicillium* fungi, in 1928,[8] the most famous example of a biopharmaceutical from biotechnology. Now, the majority of penicillin employed worldwide is produced by this kind of method in China.

Today's biotechnology in the form of genetic engineering was established by G. Mendel (1822–1884), an Austrian scientist and the father of modern genetics.[9] W. Sutton (1877–1916), an American scientist, applied the Mendelian laws of inheritance to the cellular level of living organisms and established chromosome theory.[10] The structure of DNA was discovered by J. Watson (born April 6, 1928), an American scientist, with F. Crick (1916–2004), an English scientist, in 1953. This discovery resulted in an explosion of research in molecular biology and genetics, opening the door for the biotechnology revolution.[11]

P. Berg (born June 30, 1926), an American scientist awarded the Nobel Prize in Chemistry in 1980, obtained the first recombinant DNA molecule and established the foundation for modern biotechnology.[12]

1.2 Introduction to Chemical Biotechnology and Bioengineering

1.2.1 Chemical Biotechnology and Bioengineering

Biochemistry is a traditional discipline focused on the knowledge and principles of the behavior of natural and endogenous chemicals or substances in life and biological systems. However, chemical biology is a young scientific discipline spanning chemistry, biology, and physics; it mainly uses chemistry, *i.e.* exogenous (or exogenously added) chemicals as a perturbation methodology to reveal biological laws or solve biological problems. It involves the application of chemical techniques, tools, and analyses, and chemicals from nature or produced through synthetic chemistry, for the study and manipulation of biological systems.

Science is the foundation of technology and engineering. Technology and engineering are the derivation and application of science, which really solve the practical problems related with social and natural crises. Accordingly, chemistry has chemical technology and engineering as its partner, and biochemistry has biochemical technology and engineering as its partner, so chemical biology should have its own partner—chemical biotechnology and bioengineering!

However, new methodology is needed to solve many practical problems. We know that biochemical technology or engineering can be defined as using biological methodologies and substances to solve practical and chemical problems in the areas of industry, medicine, and agriculture. Therefore, chemical biotechnology or bioengineering can be defined as using chemical methodologies and substances to solve practical and biological problems. Although one or two publications have used "chemical biotechnology or bioengineering" to describe some processes in the former case for solving chemical problems in the chemical industry, in fact these are still biochemical technology or engineering, not real chemical biotechnology or bioengineering for the processes in the latter case.

If some exogenous (or exogenously added) chemical compounds are able to regulate (enhance or attenuate) a bioprocess for some desired objectives, we believe that the whole process belongs to the area of "chemical biology" or "chemical biotechnology". However, there are some differences between chemical biology and chemical biotechnology or bioengineering. The former focuses on the theory, mechanisms, and activities in the laboratory, and the latter focuses on the operation value in the laboratory and applications in practice even beyond the laboratory.

Today, there is a distinct definition: Biotechnologies are processes that seek to transform biological materials of animal, vegetable, microbial, or viral origin into products of commercial, economic, social, and/or hygienic utility and value, and bioengineering focuses on their scale up methodology.

Therefore, the definition of chemical biotechnology and bioengineering is utilizing small chemical molecules to affect some specific bioprocesses in

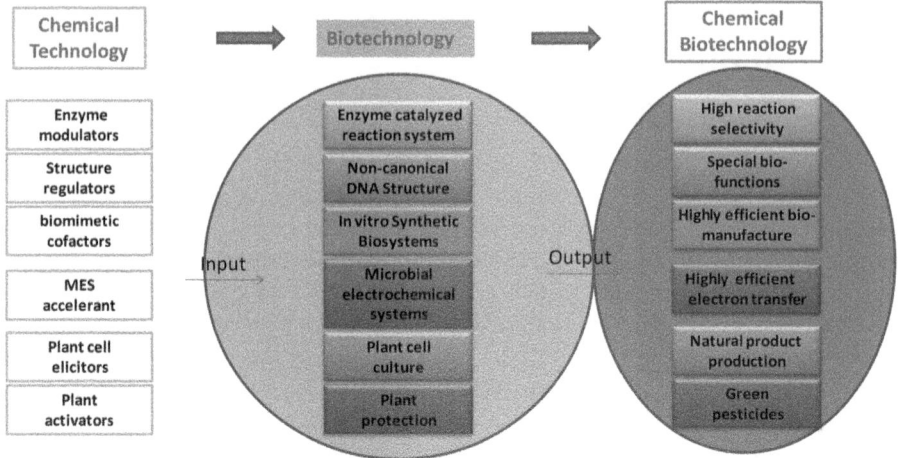

Figure 1.1   Illustration of chemical biotechnology and bioengineering.

order to make this bioprocess perform better, or on a larger scale for improving our lives and the health of our planet, which could transfer some ecological biotechnologies more economically and make the relevant chemistry greener (Figure 1.1).

In fact, for the latter these processes could be recognized as a type of green chemistry approach when used to solve chemical problems. For example, enzymatic reactions to achieve exquisite chemo-, regio-, and enantio-selectivities for pharmaceuticals with great sustainability in aqueous media and under mild conditions as described in Chapter 2, or multi-enzyme-based biotransformations to produce chiral compounds as drug precursors, sweet hydrogen, sugar biobatteries, and renewable chemicals under green conditions, as exemplified in Chapter 3.

1.2.2   Green Chemistry

Sustainability has become a central and focal issue for human beings today, even a political issue, but it previously did not attract much attention in human history. Strong disputes and conflicts as well as many stories at The World Climate Congress every year fully embody the world's anxiety on this problem. With conflicts between development and the environment, it seems to be impossible to reach a harmonious balance between GDP (Gross Domestic Product) and GWP (Global Warming Potential). In this situation and stage, green chemistry is a new and acceptable solution or option that has been proposed to solve the balance between the economy and ecology.

Green chemistry embodies two main aspects. First, it emphasizes the efficient utilization of raw or natural materials and the concomitant elimination of waste. Second, it deals with the health, safety, and environmental issues associated with the manufacture, use, and disposal or re-use of chemicals.

Green chemistry as an important concept first appeared in the early 1990s, about 20 years ago. Since then, it has made great progress in many areas, including petrochemicals, pharmaceuticals, household products, agriculture, aerospace, automobiles, cosmetics, electronics, and energy. There are hundreds and thousands of examples of successful applications of award winning, cost-effective, or economically competitive technologies. Many of them have played a significant role in informing sustainable design.[13] Important early stories include the US Presidential Green Chemistry Challenge Awards established in 1995 and the publication of the first volume of Green Chemistry, a journal from the Royal Society of Chemistry, in 1999.[14]

Green chemistry is a very important issue to scientists, engineers and society, and biotechnology is an efficient and attractive route to make chemicals and manufacturing processes cleaner or greener, efficiently and ecologically. However, from the economical view, most natural bioprocesses are still not robust enough for practical applications and industry; some chemical inducing agents, *e.g.* elicitors, modulators, and activators, are needed to promote or enhance the efficiencies of these biotechnologies, therefore, chemically promoted biotechnology and bioengineering make sense in this area.

1.3 Some Examples of Chemical Biotechnology and Bioengineering

1.3.1 Chemical Modulators for Enzymatic Reactions

In many cases, from the perspective of chemistry, it is difficult to understand the results of enzymatic reactions. For example, it seems strange that only one enantiomer in a racemic compound reacts in an enzymatic reaction if we do not understand the specific interactions between the enzyme and the substrate at the molecular level. The special compositions and structures of biomacromolecules endow enzymes with special functions. A typical, well-modulated biotransformation is usually considered to be a green process. It is estimated that biocatalysis technologies will decrease consumption of raw materials, water resources, and energy, and reduce waste emissions by 30% in 2020.[15]

Most enzymatic reactions are performed under mild conditions, such as in water, at room temperature, and at atmospheric pressure, leading to lower energy consumption and operational risks. In addition, it is widely known that enzymes are capable of catalyzing highly regio-, chemo-, and enantioselective reactions without performing the lengthy chemical protection/deprotection steps required in traditional chemical synthesis.

In spite of the superiority mentioned above, biotransformations are usually not satisfactory in industry without being carefully modulated. One dominant advantage, and sometimes also a drawback, encountered in biotransformations is that enzymatic reactions are traditionally performed in aqueous environments. However, the majority of organic chemicals are water-immiscible, resulting in very low substrate loading and low volumetric productivity.

In order to enhance the efficiency of biotransformations, chemical modulation can be used, on a case by case basis, to tune the enzyme activity, selectivity, and stability under specific reaction conditions.[16] General methods could be used to improve reaction efficiency, for example, the selection of solvent systems, buffer salts, certain metal ions, and pH adjustments have been reported to work well. The structures and properties of enzymes can be readily regulated in the presence of small-molecule modulators. For example, it is well known that enzymes are able to coordinate with metal ions, which leads to some changes in their properties. In this context, the design and application of chemical modulators in enzymatic reactions is a typical example of chemical biotechnology, as detailed in Chapter 2.

1.3.2   Chemical Biotechnology in the Regulation of Non-canonical DNA Structures

The regular double-helix structure of DNA, typically B-DNA, in which two complementary strands are held together by Watson–Crick base pairs, is well recognized. Recently it has been found that under certain conditions DNA can form non-canonical structures, such as Z-DNA, A-motif, tetraplex, triplex, hairpin, and cruciform. These structures are particularly seen in the human genome with repeat DNA sequences, and some of them have been proposed to participate in several biologically important processes, including gene regulation, expression, and evolution, and thus could be potential drug targets.

The structures and properties of these non-canonical DNA are closely related to their biological functions. Due to their unique three-dimensional structures, small molecules can bind to them to stabilize or alter their structures, and are eventually able to regulate their biological functions. One of the most successful such small molecules is *cis*-diamminedichloroplatinum(ii), or *cis*platin, a commonly used anti-cancer drug. It can covalently bind to DNA molecules, forming a DNA–*cis*platin adduct that eventually inhibits DNA synthesis. Therefore, exploring the small molecules that can interact with DNA, especially with non-canonical DNA molecules, is an effective route to anti-cancer drug discovery.

In addition to small molecules, carbon materials such as carbon nanotubes (CNTs) and graphene oxides (GO) also exhibit the ability to tune the structure of typical helical DNA and non-canonical DNA structures due to their unique structural, chemical, and physical properties. Thus, their interactions with DNA attract tremendous research interest from scientists from different fields. Particular focus will be given to the applications of CNTs and GO in gene delivery and anti-cancer drugs in Chapter 3.

1.3.3 *In Vitro* Chemical Biotechnology Biosystems for Biomanufacturing

The use of single enzymes has been commercialized for more than a half century for the production of fructose, chemicals, and semi-synthetic antibiotics. The use of cell extracts for the production of high-value vaccines, vitamins, and proteins has been studied for the last two decades. Whole cells, especially microbes, have been utilized in the production of fermented food, beer, wines, drugs, chemicals, and so forth, to meet mankind's myriad needs for thousands of years.

In vitro biosystems, also called synthetic pathway biotransformations, synthetic chemistry methodology approaches, enzyme cocktails, synthetic cascade enzyme factories, synthetic cascade manufacturing, synthetic biochemistry, and so on, are the *in vitro* assembly of a number of enzymes, which may be isolated from different organisms, and/or natural or biomimetic coenzymes, for the production of desired products that may not be produced by microbes or abiotic catalysts.[17] For example, non-food cellulose can be converted to synthetic starch catalyzed by cascade enzymes in an aqueous solution requiring neither energy input nor chemical consumption.

In vitro biosystems for biomanufacturing feature several industrial production advantages over whole-cell-based biomanufacturing. High product yield is accomplished by the elimination of side reactions and no synthesis of cell mass; fast volumetric productivity can be achieved due to the better mass transfer without the barrier of cell membranes; easy product separation can be achieved without cell membranes; enzymes usually tolerate toxins and solvents much better than whole cells because of a lack of labile cell membranes; the reconstitution of synthetic enzymatic pathways can implement some non-natural reactions that could never occur in living cells; the reaction equilibrium may be shifted in favor of the product formation through well-designed synthetic enzymatic pathways.[18]

In this system, a cofactor is a non-protein chemical compound that is required for the enzyme's biological activity. It can be considered as a "helper molecule" that assists the biochemical transformations, which we can regard as one kind of chemical biotechnology.

Organic cofactors include nicotinamide adenine dinucleotide phosphate (NADP), nicotinamide adenine dinucleotide (NAD), flavin adenine dinucleotide (FAD), adenosine triphosphate (ATP), quinone compounds, and coenzyme A (CoA). Some organic cofactors are not stable enough for long-time bioprocessing. The most promising solution is the replacement of organic cofactors with low-cost and stable biomimetic ones. Such biomimetic cofactors that share a similar structure and function with their naturally occurring counterparts can be chemically synthesized at low costs and have enhanced stability.

1.3.4 Chemical Bioengineering in Microbial Electrochemical Systems

Electrochemical systems catalyzed with whole-cell microorganisms, which are termed as microbial electrochemical systems (MES), including microbial fuel cells (MFC), microbial electrolysis cells (MEC), and microbial electrosynthesis cells (MESy), provide fascinating solutions for the sustainable development of the earth. They have been demonstrated to be promising for wastewater treatment, bioenergy harvesting, CO_2 fixation and biotransformation, *etc.*, but low power output due to low efficiency of electron releasing, extracellular electron transfer, and cell–electrode interactions largely limits the practical applications of MES. Therefore, multidisciplinary efforts have been made to improve the performance of MES. Most impressively, chemical bioengineering, referring to the use of chemical strategies to manipulate biological processes, has contributed largely to the recent advances in MES technology.

In particular, redox chemicals (serving as electron shuttles) naturally synthesized by bacteria or exogenously added synthetic molecules have been proved to be directly involved in promoting extracellular electron transfer between the cells and the electrode. Moreover, electrode modification with conductive polymers or carbon nanomaterials showed great potential for the enhancement of nanoscale topological interactions and hence the extracellular electron transfer between the cells and the electrode.[19,20] Extracellular electron transfer manipulation (a microbial process) with chemical electron shuttles or electrode modifiers can be considered as a typical application of chemical bioengineering.

Regulation of cell physiology with chemical strategies is another interesting application of chemical bioengineering. Consequently, MES performance improvement was also achieved by cell physiology manipulation with chemical strategies. Cell permeability and cell adherence, which play important roles in the extracellular electron transfer and energy efficiency of MES, were successfully manipulated with the addition of surfactants or metal ions, or cell immobilization. In addition, quorum sensing signaling molecules that can coordinate the bacterial behaviors/physiology at the population level showed great promise on MES manipulation.[21]

In this context, MES manipulation with chemical electron shuttles, electrode modifiers, surfactants, metal ions, cell immobilization, and quorum sensing will be summarized to illustrate chemical bioengineering in MES and will be described in this chapter.

1.3.5 Chemical Bioengineering in Plant Cell Culture

Plant cell culture originating from Cell Totipotency Theory was proposed by Haberlandt, a German botanist, in 1902.[22] Plant cell secondary metabolites are widely used, have significant economic value, and can be made into medicines such as paclitaxel, ginsenosides, and artemisinin. The structures of some secondary metabolites are too complex to be artificially synthesized at scale for practical purpose. For example, paclitaxel, a widely used anticancer drug, is obtained by semisynthesis by Baccatin III, isolated from the bark of the pacific yew, *Taxus brevifolia*. Because of environmental and resource factors, plant cell culture is a promising alternative technology for the mass production of valuable secondary metabolites. However, in general, the production of secondary metabolites in cell culture is too low, so some manipulative techniques are necessary to promote the productivity. Among these methods, chemical elicitations (by chemical elicitors) have been one of the best approaches for dramatically increasing secondary metabolite yields.[23] The process is typical chemical biotechnology or bioengineering, in which some compounds are used to regulate a series of biological changes in cells or enzyme activity, which increases the production of secondary metabolites on a large scale.

Jasmonic acid (JA) and its methyl ester (MJA) are important members of the family of natural jasmonates. Exogenously adding MJA was shown to increase the production of secondary metabolites in a variety of plant species. Some synthetic elicitors have also been proven to be too. We describe a series of synthetic cell culture elicitors in Chapter 4, including MJA derivatives and benzothiadiazole (BTH) derivatives.[24,25] Some of them display more potent activity in *Taxus chinensis* cell culture and *Panax notoginseng* cell culture than MJA, which reflects the full application of chemical bioengineering in plant cell culture.

1.3.6 Chemical Biotechnology for Plant Protection

Agrichemicals, including pesticides, fungicides, herbicides, and rodenticides, are very effective for crop protection. However, the excessive use of agrichemicals has caused some serious problems to the environment. The suffix "cide" means kill, so the action mechanisms of these kinds of plant protectors are to kill these pests, fungi and weeds, whilst not affecting the growth of plants. Therefore, the requirement for high selectivity, low toxicity, and low residues makes pesticides development difficult. The great biotechnology advantages from genetically modified (GM) crops haven't been widely accepted due to some ethical problems and some potential concerns resulting from DNA recombination.

Like health products for people, some chemicals (plant activators) can be used to initiate systemic acquired resistance (SAR) to protect plants from a broad spectrum of diseases and pests by naturally influencing gene expression or adjusting some cascade changes including metabolism or pathogen-related (PR) protein expressions in plants, which is different from traditional pesticides or their metabolites acting directly on target insects, fungi and weeds. No antimicrobial activity *in vitro* either by the chemical itself or by its possible metabolites is the preliminary condition, so plant activators are called green plant protecting products. The process belongs to a kind of chemical biotechnology for special action mechanisms.

In Chapter 6, a systemic discussion of plant activators will be presented, including their history, action mechanisms, current situation, a few synthetic plant activators, and future developments.

References

1. J. H. Hulse, *Trends Food Sci. Technol.*, 2004, **15**, 3–18.
2. N. Qureshi, *Plant Sci. Rev.*, 2010, **2011**, 249.
3. M. G. Fári and U. P. Kralovánszky, *Int. J. Hortic. Sci.*, 2006, **12**, 9–12.
4. A. Nasim, *Commission on Science and Technology for Sustainable Development in the South*, 2003, p. 19.
5. Y. Xu, D. Wang, W. L. Fan, X. Q. Mu and J. Chen, in *Biotechnology in China II*, Springer, 2010, pp. 189–233.
6. R. Bud, *The Uses of Life: A History of Biotechnology*, Cambridge University Press, 1994.
7. C. A. Janeway, *Cold Spring Harbor Symposia on Quantitative Biology*, 1989.
8. J. W. Bennett and K.-T. Chung, *Adv. Appl. Microbiol.*, 2001, **49**, 163–184.
9. G. E. Allen, *Endeavour*, 2003, **27**, 63–68.
10. S. Y. Kim and K. E. Irving, *Sci. Educ.*, 2010, **19**, 187–215.
11. F. Crick and J. Watson, *Nature*, 1953, **171**, 737–738.
12. M. Chamberlin and P. Berg, *J. Mol. Biol.*, 1964, **8**, 297–313.
13. P. Anastas and N. Eghbali, *Chem. Soc. Rev.*, 2010, **39**, 301–312.
14. J. H. Clark, *Green Chem.*, 2006, **8**, 17–21.
15. M. Gavrilescu and Y. Chisti, *Biotechnol. Adv.*, 2005, **23**, 471–499.
16. R. C. Rodrigues, C. Ortiz, Á. Berenguer-Murcia, R. Torres and R. Fernández-Lafuente, *Chem. Soc. Rev.*, 2013, **42**, 6290–6307.
17. Y.-H. P. Zhang, S. Myung, C. You, Z. Zhu and J. A. Rollin, *J. Mater. Chem.*, 2011, **21**, 18877–18886.
18. C. You and Y.-H. P. Zhang, *ACS Synth. Biol.*, 2012, **2**, 102–110.
19. Y. C. Yong, X. C. Dong, M. B. Chan-Park, H. Song and P. Chen, *ACS Nano*, 2012, **6**, 2394–2400.
20. Y. C. Yong, Y. Y. Yu, X. H. Zhang and H. Song, *Angew. Chem., Int. Ed.*, 2014, **53**, 4480–4483.
21. Y. C. Yong, Y. Y. Yu, C. M. Li, J. J. Zhong and H. Song, *Biosens. Bioelectron.*, 2011, **30**, 87–92.
22. E. Höxtermann, *Physiol. Plant.*, 1997, **100**, 716–728.
23. H. Dörnenburg and D. Knorr, *Enzyme Microb. Technol.*, 1995, **17**, 674–684.
24. Z. G. Qian, Z. J. Zhao, Y. Xu, X. Qian and J. J. Zhong, *Biotechnol. Bioeng.*, 2004, **86**, 809–816.
25. Y. Xu, Z. Zhao, X. Qian, Z. Qian, W. Tian and J. Zhong, *J. Agric. Food Chem.*, 2006, **54**, 8793–8798.

CHAPTER 2

Chemical Modulators for Enzymatic Reactions

JIAN-HE XU*[a], GAO-WEI ZHENG[a], AND XIAO-JING LUO[a]

[a]Laboratory of Biocatalysis and Synthetic Biotechnology, State Key Laboratory of Bioreactor Engineering, East China University of Science and Technology, Shanghai 200237, China
*E-mail: jianhexu@ecust.edu.cn

2.1 Asymmetric Synthesis Using Biocatalysts

2.1.1 An Introduction to Asymmetric Synthesis

A molecule is chiral if it has a non-superimposable mirror image (Figure 2.1). The most common cause of chirality in molecules is the presence of asymmetric carbon atom(s). The majority of pharmaceuticals and natural products are in single-enantiomer form. Asymmetric synthesis plays a key role in the preparation of these compounds of interest. Stereoselective synthesis of chiral compounds, as defined by IUPAC, is a chemical reaction in which one or more new elements of chirality are formed in a substrate molecule, producing the stereoisomeric products in unequal amounts. Asymmetric synthesis is particularly important in the fields of pharmaceuticals and agrochemicals, because the different enantiomers or diastereomers of a molecule often have different biological activities.

Chirality is becoming increasingly common in pharmaceuticals, as well as in the fine chemical industry. As reported, as many as 80% of the

RSC Green Chemistry No. 34
Chemical Biotechnology and Bioengineering
By Xuhong Qian, Zhenjiang Zhao, Yufang Xu, Jianhe Xu, Y.-H. Percival Zhang, Jingyan Zhang, Yangchun Yong, and Fengxian Hu

small-molecule drugs approved by the FDA in 2006 were chiral and 75% were single enantiomers/diastereoisomers,[2] which contain an average of two chiral centers.[3] According to the regulatory requirements, a stereoisomeric purity of 99.5% is mandatory for chiral drugs.[3]

The approaches to preparing a single enantiomer or diastereomer of a chiral molecule include:

   i) Enantioselective catalysis using the coordination complexes of chiral ligands with metals;
  ii) Enantioselective organocatalysis using non-metal small organic molecules like proline;
 iii) Biocatalysis using isolated enzymes or whole cells;
  iv) Use of chiral auxiliaries; and
   v) Synthesis by using chiral pools.

Apart from enantioselective synthesis, optically pure compounds can be obtained by chiral resolution. This involves the isolation of one enantiomer from a racemic mixture by a number of methods. This route remains cost-effective when the time and money required for making racemic mixtures are low, or both enantiomers have a separate use.

As listed in Table 2.1, biocatalytic technology employing either isolated enzymes or whole-cell biocatalysts is gaining popularity due to its superior chemoselectivity, regioselectivity, or stereoselectivity, as well as mild reaction conditions and environmental biocompatibility.

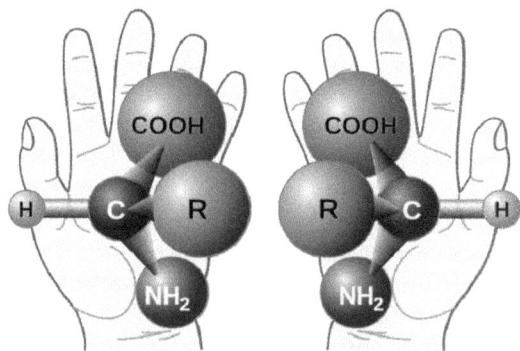

Figure 2.1   Graphical illustration of two enantiomers of an amino acid molecule.[1]

Table 2.1   Percent contribution of various technologies to chiral chemical manufacturing.

Year	Chiral pools and resolution	Chemical catalysis	Biological catalysis
2002	55%	35%	10%
2005	49%	36%	15%
2009	41%	36%	23%

2.1.2   Enzyme Classification and Examples

Based on the types of reactions that they catalyze, enzymes are generally classified into six major categories (Table 2.2). It has been estimated that about 60% of current biotransformations rely on hydrolases, followed by 20% on oxidoreductases. Moreover, some C–C bond forming and oxygenation enzymes catalyze reactions with very high efficiency and very low waste generation, highlighting the potential of emerging enzymes.

2.1.3   An Overview of Enzymes Used in Chiral Synthesis

Though asymmetric hydrogenation,[4] isomerization,[5] and epoxidation[6] have been well developed for the synthesis of enantiopure compounds, the universality of these methods is limited. Classical resolution of a racemate with a chiral auxiliary is typically used by chemists for manufacturing optically active compounds, although these methods suffer from low efficiency and large amounts of waste. More importantly, the limitations are aggravated by the unsatisfactory optical purity of the products.

Taking advantage of several aspects, biocatalysis is a robust competitor and a good alternative to chemical processes for the preparation of enantiopure compounds due to the intrinsic exquisite stereo- or regio-selectivity of enzymes under mild reaction conditions, such as in aqueous medium, at physiological pH, and at ambient temperature. Activation and protection/deprotection steps of functional groups are generally not required for biotransformations. Interestingly, the complex chemical synthesis process for products with multiple chiral centers may be resolved by a simple and elegant bioconversion process.[7–9] The concise bioprocess steps make biocatalysis more attractive for efficient production with less waste, *i.e.*, reducing the *E*-factor.[7]

Classical kinetic resolution of racemates is frequently employed for the preparation of enantiopure compounds. In order to circumvent the limitation

Table 2.2   Enzyme classification and examples.

Enzyme class	Reaction catalyzed	Examples
Hydrolases	Hydrolytic reactions in H_2O	Lipase, protease, esterase, nitrilase, amidase, glycosidase, phosphatase
Oxidoreductases	Oxidation or reduction	Dehydrogenase, oxidase, oxygenase, peroxidase
Transferases	Transferring a group from one molecule to another	Transaminase, transaldolase, glycosyltransferase
Lyases	Non-hydrolytic bond cleavage	Decarboxylase, dehydratase, deoxyribose-phosphate aldolase
Isomerases	Intramolecular rearrangement	Racemase, mutase
Ligases	Bond formation requiring triphosphate	DNA-ligase

of 50% yield in classical bioresolution, dynamic kinetic resolution processes based on *in situ* enzymatic or transition metal-catalyzed substrate racemization were developed.[10–12] Compared with enantiomeric resolution, asymmetric biotransformations are more attractive because enantiopure compounds can be obtained in theoretical yields of up to 100% with excellent atom efficiency. Examples of biocatalyzed asymmetric synthesis are listed in Table 2.3. Among them, asymmetric carbonyl reduction by dehydrogenases, stereoselective

Table 2.3 Biocatalyzed asymmetric syntheses.

Substrate	Enantiopure compound	Enzyme	Reaction type
Ketone	OH on R–R'	Dehydrogenase	Reduction
Enoate	O, R' on R with R"	Enoate reductase	Enoate reduction
Sulfide	R–S(O)–R'	Monooxygenase	Sulfoxidation
Alkene	epoxide R–R'	Monooxygenase haloperoxidase	Epoxidation
Olefin or aromatic ring	HO, OH on R–R'	Dioxygenase	Dioxygenation
Alkane or benzyl	OH on R, R', R"	Monooxygenase	Hydroxylation
Ketone	lactone R–O	Monooxygenase	Baeyer–Villiger oxidation
Alkene	OH on R–R', Halo	Haloperoxidase	Halohydrin formation
Ketone, aldehyde	OH on R–CN, R'	Hydroxynitrile lyase	Hydroxynitrilation
α-Keto acid	NH₂ on R–COOH	Transaminase	Transamination

hydroxylation by monooxygenases, and asymmetric hydroxyl nitrilation by hydroxynitrile lyases have become the topics of common interests.

Industrial production places high demands on biocatalysis and biotransformation.[13] Poor enzyme stability, low solubility of hydrophobic substrates in the aqueous phase, and low regeneration efficiency of cofactors are the bottlenecks to the application of bioprocesses application. However, the efficiency (cost, yield, and productivity) of biocatalytic processes has improved significantly with the development in biotechnology techniques, such as immobilization of biocatalysts (enzymes or whole cells), introduction of two-phase reaction systems, and cofactor regeneration by whole-cell catalysis or enzyme-coupled systems. Most importantly, biocatalysis is limited by the scarcity of biocatalysts with excellent properties. Developed with breakthroughs in biotechnology, for example, in molecular biology, genomic sequencing and bioinformatics, an increasing spectrum of biocatalysts for industrial application is being readily discovered by enzymologists through approaches such as data mining, directed screening, or molecular evolution.[14]

With the access to diverse and stable biocatalysts, more and more conventional chemical processes (first generation) in pharmaceutical manufacturing have been replaced by second-generation biocatalysis processes with substantial impact on the pharmaceutical industry.[15] In this chapter, some commonly used biocatalytic reactions for chiral preparation, including hydrolytic reactions, acyl and glycosyl transfer reactions, asymmetric reduction/oxidation reactions, and asymmetric formation of C–C bonds, are introduced and exemplified by the research achievements developed by the authors' laboratory as well as other research groups. Some of the bioprocesses described herein have been successfully applied on pilot or even industrial scale.[16–18]

2.2   Chemical Modulators for Enantioselective Biocatalysis

In industrial biotransformations, hydrolytic reactions occupy a prominent position for the production of optically active amines, alcohols, and carboxylic acids.[19] Compared with other reactions, hydrolytic reactions are feasible to scale up because they are cofactor-free, relatively simple, and chemically tunable systems. In addition to home-made whole-cell biocatalysts, which are considered to be more cost-effective for specific syntheses, some commercially available hydrolases, including lipases/esterases, epoxide hydrolases, nitrilases, and glycosidases, are also employed for the enantioselective production of chiral chemicals.

2.2.1   Enantioselective Hydrolysis of Ketoprofen Esters Using Yeast Cells

α-Arylpropionic acids, such as naproxen, ibuprofen, ketoprofen, flurbiprofen, and suprofen, are a class of anti-inflammatory drugs. Their anti-inflammation activities are dominated by the (S)-isomers, so the production

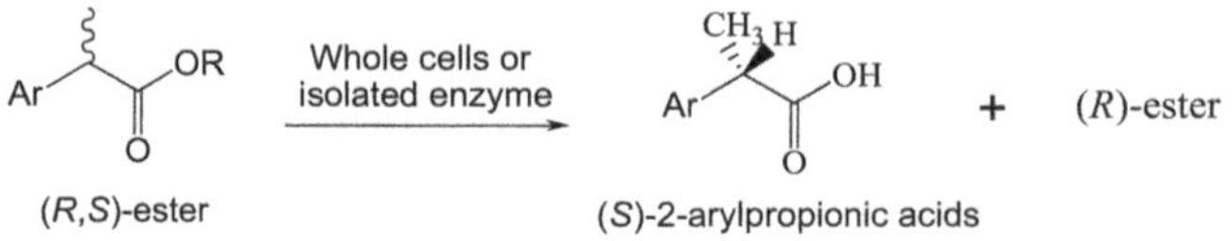

Scheme 2.1 Enantioselective hydrolysis of α-arylpropionic acid esters using whole cells or isolated enzymes.

of (S)-α-arylpropionic acids has been a research hotspot and well developed in the past decades. Enzymatic hydrolysis of racemic esters has now become a standard and easy-to-modulate route for preparing (S)-α-arylpropionic acids.

Ketoprofen, or 2-(3-benzoylphenyl)propionic acid, is a kind of 2-arylpropionic acid that represents an important group of nonsteroidal anti-inflammatory drugs (NSAIDs), and is widely used to relieve inflammation and pain resulting from arthritis, sunburn, menstrual pain, or fever. Previous studies have shown that its anti-inflammatory activity lies mainly in its (S)-enantiomer, while its (R)-enantiomer is a potentially highly potent analgesic treatment for neuropathic pain. Therefore, it is very valuable to obtain optically pure (S)- or (R)-ketoprofen.

A yeast strain, *Trichosporon brassicae* CGMCC 0574, was identified from 92 strains of soil isolates for its high (S)-selectivity in the hydrolysis of ketoprofen ethyl ester. The effective strains of the microorganisms were isolated from soil samples with the ester as the sole carbon source. The ethyl ester proved to be the best substrate for resolution of ketoprofen among several ketoprofen esters examined. The hydrolysis of ketoprofen ethyl ester (Scheme 2.1) driven by the resting cells of *T. brassicae* CGMCC 0574 gave (S)-ketoprofen in an enantiomeric excess (ee) of 91.5% at 42% conversion, exhibiting a moderate enantioselectivity ($E = 45$) that needs to be further enhanced.[20]

2.2.1.1   *Effects of Chemical Solvents on Whole-Cell Activity*

Although the catalytic activity of *T. brassicae* CGMCC0574 whole cells in the enantioselective hydrolysis of ketoprofen ester was initially moderate, it was significantly promoted after the cells were pretreated for a few hours in a buffer containing an alcohol as the modulator.[21] This suggested a possible permeability barrier from the cell membrane against the mass transfer of substrates and/or products. Therefore, the permeabilization of the yeast cells was considered to be favorable for the enzymatic resolution of ketoprofen.

A suitable permeabilizing agent depends on the organism and the composition of the cell membrane. Many investigators have employed toluene as a permeabilizing agent. However, in this study, a 30% decrease in the cell activity was observed after toluene treatment. Whereas, among several other reagents tested (Table 2.4), acetone and 2-propanol treatment raised the whole-cell activity by 161% and 235%, respectively, relative to the original activity, indicating their potency in permeabilizing the yeast cells.

Meanwhile, the total protein concentration in the supernatant of the treated cell suspension was very low, and no notable activity was detected toward the ethyl ester of ketoprofen. Therefore, it is safe to conclude that

Table 2.4 Activity and enantioselectivity of *T. brassicae* cells after pretreatment with various organic solvents.[21]

Organic solvent	Relative activity (%)	Enantiomeric ratio
Control	92	54
Toluene	67	N.D.
Ethanol	79	N.D.
2-Propanol	235	56
Acetone	161	51
DMSO	111	N.D.

the treatment of whole cells with organic solvents did not lead to cell disruption and that the pretreatment process did not cause leakage of the requisite enzyme from the yeast cells.

As for the enantioselectivity of the cells treated with acetone or 2-propanol, it was shown that the treatment with these two solvents did not impair the enantioselectivity of the cells toward ethyl (*S*)-ketoprofenate. Finally, 2-propanol was chosen as the best permeabilizing agent for further study owing to the remarkably higher activity than that of acetone treatment.

2.2.1.2 Modulating the Whole-Cell Mediated Resolution with Additives

As mentioned above, ketoprofen is widely used in clinical practice as a non-steroidal anti-inflammatory drug, like the other 2-arylpropionic acids such as naproxen and ibuprofen. The anti-inflammatory activity of ketoprofen was previously believed to reside in its (*S*)-enantiomer. However, research has indicated that the (*R*)-enantiomer of ketoprofen was able to prevent periodontal disease and was thus of pharmacological value as a toothpaste additive. It has also been discovered that (*R*)-ketoprofen has several previously unappreciated advantages as an analgesic and antipyretic. Another yeast strain, *Citeromyces matriensis* CGMCC 0573, was therefore isolated for the enantioselective hydrolysis of (*R*)-ketoprofen ethyl ester.[22]

The effect of some selected additives on the hydrolysis reaction was first examined. As shown in Figure 2.2, the additives significantly increased both the ee_p value and the conversion ratio. The conversion was enhanced by two-fold using ethanol as an additive, and further increased by about seven-fold in the presence of Tween-80. Most likely, the greater dispersion of the insoluble ketoprofen ester by the additives leads to an improved diffusion rate and larger interface area between the substrate and the cells, both of which are beneficial for improving the catalytic efficiency of the resting cells of *C. matriensis* CGMCC 0573.

The initial rate of ketoprofen ester hydrolysis in a reaction mixture containing 0.5% (w/v) Tween-80 was examined at temperatures ranging from 20 to 50 °C. The optimum temperature for the activity of resting cells was 40 °C, while the activity decreased sharply at higher temperatures. The activity at 50 °C was only 14% of that at 40 °C; however, 30 °C was the optimum temperature for the highest enantioselectivity.

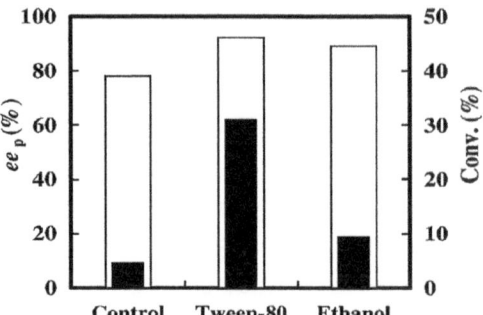

Figure 2.2 Modulating effects of additives on the resting cells of *Citeromyces matriensis* CGMCC 0573. Reactions were carried out in 10 mL of potassium phosphate buffer (0.1 M, pH 7.0) containing 10 mM ester and either 0.5% (w/v) Tween-80 or 5% (v/v) ethanol for 16 h. (■) Conversion; (□) ee_p.

The activity was higher under a slightly alkaline environment (pH 8.0) in 0.1 M citrate/0.2 M Na_2HPO_4 buffer, which is very different from that reported for *Candida rugosa* lipase, which expressed much higher activity and enantioselectivity under an extremely acidic environment (pH 2–4).[23] On the other hand, there was no evident change in the ee_p values of ketoprofen produced between pH 4.0 and 10.0. The enzyme was most stable in buffers with a pH ranging from 5.0 to 7.0, retaining more than 60% of the initial activity after incubation at 30 °C for 12 h. Based on these results, pH 7.0 and 30 °C were chosen as the optimal conditions for the yeast-cell mediated reaction.

As mentioned above, the addition of the non-ionic surfactant Tween-80 resulted in a significant enhancement of the enzyme activity and enantioselectivity during the ester hydrolysis process. Therefore, the effect of Tween-80 concentration on the reaction was further investigated using resting cells at pH 7.0 and 30 °C. It was found that both the conversion ratio of the substrate and the ee_p were enhanced with increasing load of Tween-80 below 0.5% (w/v), but they tended to stay constant beyond 0.5% load. Thus, the optimal concentration of Tween-80 was 0.5% (w/v) in terms of the conversion, which was enhanced by over 7-fold.[22]

2.2.2 Enzymatic Hydrolysis of Chiral Esters Using Lipase or Esterase

Lipase- or esterase-catalyzed hydrolytic resolution of ketoprofen esters was usually employed for the preparation of (*S*)-ketoprofen, as shown in Scheme 2.1. Either isolated microorganisms[20–22] or commercial lipases[23,24] have been applied for the enantioselective hydrolysis of racemic ketoprofen esters. A commercial enzyme, Lipase OF from *Candida rugosa* (Meito Sangyo, Japan), was found to be a good biocatalyst for kinetically resolving the 2-chloroethyl ester of ketoprofen (2-(3-benzoylphenyl)propionic acid).[24]

2.2.2.1 Modulating Effect of Surfactants on Candida Rugosa Lipase Catalysis

Since the ketoprofen ester is a water-insoluble substrate, the substrate dispersion should be a critical factor affecting the bioconversion reaction. It was found that addition of surfactants as modulators could effectively improve the substrate dispersion and conversion, and consequently enhanced the lipase activity.[24] With the addition of 2% (w/v) Tween-80 or 3% (w/v) OP-10 (nonyl phenol polyethyleneoxy ether), the activity of the crude lipase was greatly increased by 13- and 15-fold, respectively. Interestingly, addition of the surfactant as a chemical modulator also significantly improved the enantioselectivity of the enzyme. The enantiomeric ratio (E value) was greatly enhanced from 8 to 100 with the purified lipase (L2) in the presence of 2% (w/v) Tween-80 (Figure 2.3).

With respect to the purified isoenzyme (L2) of Lipase OF in the presence of Tween-80 as a modulator, a similar effect was observed. The activity of L2 was about six times higher in the presence of 0.3% (w/v) Tween-80 than that without the surfactant. It took only 10 h to obtain a 50% conversion in the presence of 2% Tween-80, whereas the reaction proceeded to only 14% after 24 h in the absence of Tween-80. When the surfactant concentration was higher than 2%, the activity could not be further enhanced for either the purified enzyme or the crude one.

Similar to the activity, the enantioselectivity of the crude and purified enzymes also depends on the concentration of modulator Tween-80 (Figure 2.3). Adding Tween-80 from 0 to 8% (w/v), the E value of the crude lipase gradually increased from 1.2 to 6.7, while no further enhancement was observed above 8%. For the isoenzyme L2, which contained the majority of the lipase activity recovered, the E value increased along with ascending Tween-80 concentration, reaching 100 when the modulator concentration was 2% (w/v). Up to 10% (w/v) Tween-80, the E value was still maintained at a high level over 100. Hence, in the presence of 2% Tween-80, the product (S)-ketoprofen could be prepared with a fairly high optical purity of 96.4% ee

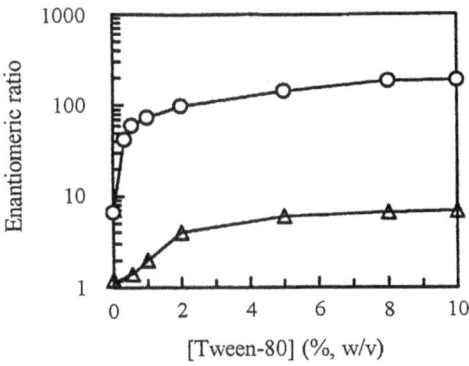

Figure 2.3   Modulating the enantioselectivities of crude lipase (Δ) and an active component L2 (○) by adjusting the Tween-80 concentration.

through the lipase L2-catalyzed hydrolysis of ketoprofen ester at 49.5% conversion. Consequently, a simple and efficient modulation method of adding Tween-80 leads to a dramatic improvement of the enzyme enantioselectivity in the hydrolysis of racemic ketoprofen ester.

2.2.2.2 *Improving Lipase Catalysis by Tuning pH or Using Ionic Resin*

The pH of the reaction medium is another important factor for modulating both the activity and enantioselectivity of Lipase OF.[23] The enzyme showed optimal hydrolysis activity at pH 4.0, while the enantioselectivity increased sharply with the decrease in medium pH from 4.0 to 2.2. Based on spectroscopic studies,[25] the enhancement of the lipase activity and enantioselectivity at the lower pH could be attributed to the changes in the flexible and sensitive conformation of the lipase induced by tuning the biocatalyst microenvironment. Using a hybrid strategy by modulating pH and surfactant, enantiomer-enriched (S)-ketoprofen could be obtained with 95.5% ee and 39.1% yield from *rac*-ketoprofen chloroethyl ester (100 mM) at pH 2.5 in the presence of 0.5% (w/v) Tween-80 as a modulator.[26]

After terminating the reaction, the pH of the reaction mixture was adjusted to *ca.* 10.0 and then centrifuged to remove the unreacted (R)-2-chloroethyl ketoprofen. The supernatant containing the sodium salt of (S)-ketoprofen was acidified to pH 2.0 to generate a precipitate of (S)-ketoprofen. The resulting product could be further refined by recrystallization (*e.g.*, in ethanol).

The crude preparation of Lipase OF contains several isoenzymes, which could be separated by column chromatography on a cation ion resin (Sephadex C-50) and eluted by citrate–phosphate buffers of pH 3.3, 4.7, and 6.8. The active fractions eluted were designated as L1, L2, and L3, respectively.[27] Active fractions L2 and L3 showed higher enantioselectivities as compared with fraction L1. Considering that the lipase has higher enantioselectivity at lower pH,[23] the partial purification of the crude enzyme was integrated with the enzyme immobilization by simply adsorbing the lipases onto Sephadex C-50 at pH 3.5.[27] Thanks to the selective removal of the unfavorable lipase isoenzyme L1, the tightly fixed enzyme components on the resin displayed significantly improved enantioselectivity, and (S)-ketoprofen was obtained with >94% ee at 22.3% conversion in the presence of 0.5 g L^{-1} Tween-80 at pH 3.5. The operational stability of the immobilized biocatalyst was examined in a packed column and air-bubbled column reactors. As a result, the air-bubbled column was an ideal bioreactor, which could be operated smoothly for at least 350 h, retaining nearly 50% of the activity of the immobilized lipase in the end.

2.2.2.3 *Combinational Strategies for Modulating Enantioselectivity*

Optically pure 1-(3′,4′-methylenedioxyphenyl)ethanol is a key chiral intermediate for the synthesis of Steganacin and Salmeterol. A *para*-nitrobenzyl esterase cloned from *Bacillus amyloliquefaciens* (BAE) was employed to hydrolyze

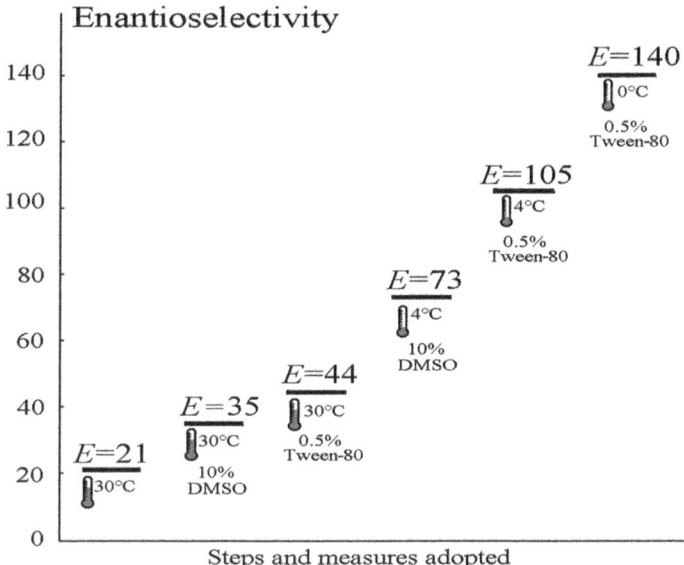

Figure 2.4 Stepwise and combinational strategies for modulating the enantiose-lectivity of a *para*-nitrobenzyl esterase cloned from *Bacillus amylolique-faciens* (BAE) in the hydrolysis of 1-(3′,4′-methylenedioxyphenyl)ethyl acetate.[28]

1-(3′,4′-methylenedioxyphenyl)ethyl ester to produce (*R*)-1-(3′,4′-methylene-dioxyphenyl)ethanol.[28] Initially, only moderate enantioselectivity (*E* = 35) was obtained at 30 °C. In order to modulate the enantioselectivity of BAE-catalyzed reaction, some strategies such as alteration of the reaction tem-perature and introduction of additives were tried (Figure 2.4). It was shown that the enantioselectivity (*E*-value) of BAE was significantly improved up to 140 by adding Tween-80 and lowering the reaction temperature to 0 °C. The result was confirmed in a decagram-scale preparative bioresolution. These results provide a more selective and tunable process for the bioproduction of (*R*)-1-(3′,4′-methylenedioxyphenyl)ethanol and a general strategy for mod-ulating the stereochemical selectivity of an enzyme for synthetic purposes.

2.2.3 Enantioselective Hydrolysis with Epoxide Hydrolases

Enantiopure epoxides and their corresponding vicinal diols are extensively employed as chiral building blocks for the synthesis of various bioactive products in the pharmaceutical and agrochemical industries due to their high ability to react with a broad variety of nucleophiles. Therefore, develop-ing enzymes and modulation methods is of great interest for the biosynthe-sis of enantiopure epoxides or diols. One of the most promising approaches to synthesize such chiral synthons under environmentally gentle conditions is the enantioselective hydrolysis of racemic epoxides (Scheme 2.2) using cofactor-independent epoxide hydrolases (EHs) [EC 3.3.2.X].

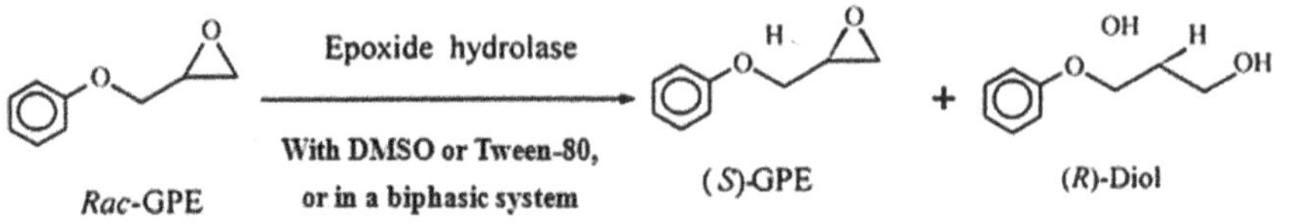

Scheme 2.2   Biocatalytic resolution of racemic glycidyl phenyl ether using BmEH under modulated conditions.

Phenyl glycidyl ether (PGE) is a potentially useful compound for the synthesis of chiral amino alcohols and bioactive compounds such as β-blockers. An isolated bacterial strain, *Bacillus megaterium* ECU1001, is capable of utilizing phenyl glycidyl ether as the sole carbon source and energy source.[29] The epoxide hydrolase from *B. megaterium* ECU1001 (BmEH) was biosynthesized in parallel with cell growth and reached a maximum activity of 31.0 U L^{-1} after 30 h of culture when the biomass was 9.1 g$_{DCW}$ L^{-1}. The lyophilized whole cells of *B. megaterium* could preferentially hydrolyze the (*R*)-glycidyl phenyl ether at 35 °C and pH 8.0, generating (*S*)-epoxide and (*R*)-diol with a moderate enantioselectivity (*E* = 47.8). The (*S*)-epoxide remained in the reaction mixture with >99.5% ee and a yield of 44.1%.

2.2.3.1   Modulating the Activity with Cosolvents or Surfactants

The addition of an organic cosolvent or a surfactant as a chemical modulator was tested for improving the solubilization of the substrate (PGE) in an aqueous system. The nature of the cosolvents greatly affected the rate of epoxide hydrolysis, as shown in Table 2.5. Indeed, in the absence of the cosolvent the activity was low. The highest activity was observed when using DMSO as the cosolvent. It was previously found that Tween-80 could significantly enhance the lipase performance in the resolution of ketoprofen ester.[24] Therefore, we also examined the modulating effect of Tween-80 on the performance of *BmEH*. Interestingly, the highest activity (2.40 U mg^{-1}, 177%) was observed in the presence of Tween-80. Almost no activity of *BmEH* could be detected in the presence of tetrahydrofuran and acetone. Although similar activities were observed with DMSO and methanol, the latter was abandoned because a chemical reaction occurred spontaneously between PGE and methanol after incubation for a period of time at 30 °C.[30]

The maximum activity of the isolated enzyme was observed at 30 °C and pH 6.5 in a buffer system with 5% (v/v) DMSO as a cosolvent. The enzyme was very stable at pH 7.5 and retained full activity after incubation at 40 °C for 6 h. Interestingly, when the cosolvent DMSO was replaced by an emulsifier (Tween-80, 0.5% w/v) as an alternative modulator to disperse the water-insoluble substrate, the apparent activity of the epoxide hydrolase significantly increased by 1.8-fold, while the optimum temperature shifted from 30 to 40 °C and the half-life of the enzyme at 50 °C increased by 2.5 times (Figure 2.5). The enzymatic hydrolysis of *rac*-PGE was highly enantioselective, with an *E*-value (enantiomeric ratio) of 69.3 in the Tween-80 emulsion system, which is obviously superior than that (41.2) observed in the DMSO-modulated system.[30]

Table 2.5 Modulating effect of cosolvents on BmEH activity.[a]

Additives[b]	Specific activity (U mg^{-1})	Relative activity (%)
Control	0.17	13
DMSO	1.36	100
Methanol	1.40	103
Ethanol	0.54	40
DMF	0.39	29
Acetonitrile	0.33	24
Acetone	0.15	11
Tetrahydrofuran	0.07	5
Tetrahydrofuran	0.07	5

[a]The partially purified enzyme was incubated with 10 mM GPE in the presence of the different cosolvents in Tris–HCl buffer (10 mM, pH7.5).
[b]The final concentration of additives was 5% (v/v).

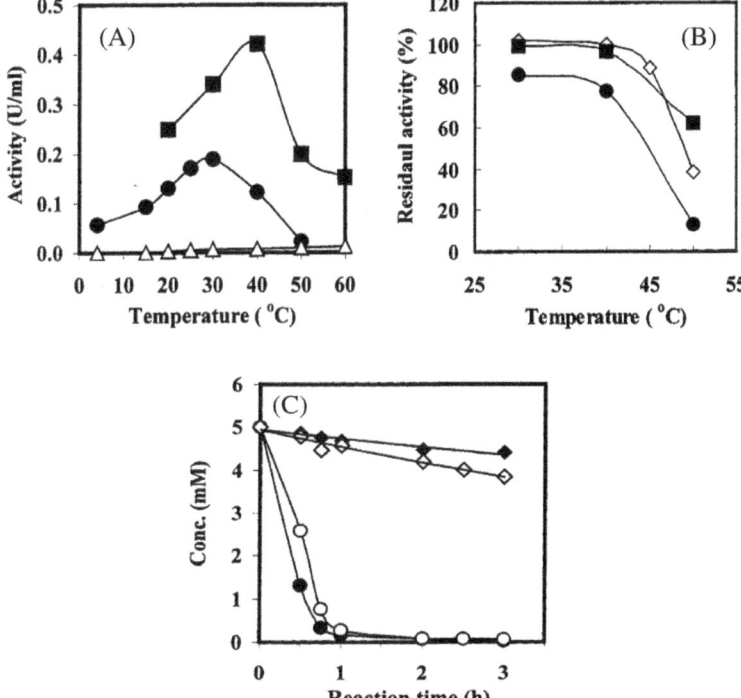

Figure 2.5 Enzymatic resolution of racemic glycidyl phenyl ether (10 mM) by partially purified enzyme in Tris–HCl buffer (pH 7.5). Part A: effect of temperature on EH activity with either (■) Tween-80 (0.5%, w/v) or (●) DMSO (5%, v/v); (Δ) represents spontaneous hydrolysis without enzyme. Part B: residual EH activity after incubation at the indicated temperatures for 1 h with (●) 5% (v/v) DMSO or (■) 0.5% (w/v) Tween-80, or (◊) without additives. Part C: time-course of epoxide enantiomers: (*R*)-GPE (●, o) and (*S*)-GPE (♦, ◊), in the presence of (o, ◊) DMSO (5%, w/v) or (●, ♦) Tween-80 (0.5%, w/v).

The gene encoding *Bm*EH, which showed unusual (*R*)-enantioselectivity and very high activity, was successfully cloned from *Bacillus megaterium* ECU1001. The highest enantioselectivity (*E* > 200) was achieved in the bioresolution of *ortho*-substituted PGE and *para*-nitrostyrene oxide.[31] It is worth noting that the substrate structure remarkably modulates the enantioselectivity of the enzyme, as a reversed (*S*)-enantiopreference was unexpectedly observed for the *ortho*-nitrophenyl glycidyl ether. As a proof-of-concept, five enantiopure epoxides (>99% ee) were obtained in high yields and a gram-scale preparation of (*S*)-*ortho*-methylphenyl glycidyl ether was successfully achieved within a few hours, indicating that *Bm*EH is an attractive and adjustable biocatalyst for the efficient preparation of optically active epoxides.

2.2.3.2   Performance of Epoxide Hydrolase in a Biphasic System

The biocatalytic resolution of epoxides has mainly been performed in single aqueous buffer systems. In our laboratory, *Bm*EH has been proved to be very useful for chiral synthesis of (*R*)-GPE on a preparative scale. More importantly, this enzyme exhibited a complementary enantiospecificity as compared with those described so far, affording the unreacted epoxide in (*S*)-configuration which is the solely useful enantiomer for synthesis of bioactive β-blockers. In a single aqueous phase, however, most epoxides including GPE may be spontaneously hydrolyzed into vicinal diols without any enantiospecificity. In addition, only at a very low concentration can epoxides be dissolved in aqueous media. The instability and low solubility of epoxides in the aqueous phase may result in a remarkable decrease in the yield of kinetic resolutions, thus limiting the application of these resolution processes on a practical scale.

In order to overcome the above drawbacks observed in single aqueous phase hydrolytic resolution, attempts have been made on nonconventional media such as an organic-aqueous two-phase system. The epoxide predominantly partitions into the organic phase, *i.e.* the water-immiscible organic phase can act as a reservoir for the unstable epoxide substrates, thus regulating the concentration of these compounds in the biocatalyst microenvironment and minimizing their toxicity. The diol products will be extracted by the aqueous phase, thereby facilitating the post-reaction process. Very few papers have been published regarding the optical resolution of epoxides in two-phase systems, although this system is relatively popular in the field of bioconversion. Herein we describe the use of an organic–aqueous two-phase system for the bioresolution of *rac*-PGE by selecting a suitable organic solvent.[32]

The partitioning behavior of the epoxide (substrate) and the diol (product) between the two phases was first investigated. As a result, both the epoxide and the diol were mostly dissolved in the organic phase when chloroform, dodecanol, or *n*-decane was used. Moreover, dodecanol and *n*-decane easily formed emulsions with the buffer (potassium phosphate or KPB), which were difficult to separate. When using *n*-hexane, *n*-octane, or isooctane as the organic solvent, the epoxide partitioned mainly in the organic phase while the diol dissolved mainly in the aqueous phase. The partition difference between the substrate and the product was the maximum when using isooctane as the organic phase. Consequently, *n*-hexane, *n*-octane, and isooctane were chosen as suitable candidates for achieving a better separation between the substrate and the product.

The second criterion for the selection of a proper organic solvent was its effects on the enzyme activity and stability. Seven organic solvents were tested, chloroform, cyclohexane, *n*-hexane, *n*-octane, isooctane, dodecanol, and *n*-decane, with log P values of 2.0, 3.2, 3.5, 4.5, 4.5, 5.0, and 5.6, respectively. The log P of a solvent, the logarithm of the partition coefficient of the solvent in a standard mixture of 1-octanol and water, is a parameter often used for predicting its biocompatibility, and is usually more indicative of the dissolved solvent.

In our experiments, the diol-producing velocities observed in the presence of the tested solvents were similar, with the exception of chloroform and cyclohexane. The specific production rate (nmol min^{-1} mg$_{DCW}^{-1}$) observed can be ranked in the following order: *n*-hexane (10.31) > isooctane (10.25) > *n*-octane (9.79) > *n*-decane (9.18) > dodecanol (8.76) > cyclohexane (5.34) > chloroform (0.64). The relative effect of the solvents correlates well with the biocompatibility criteria since the organic solvents with log P > 4.0 gave higher enzyme activity, whereas chloroform (log P = 2.0) displayed the greatest inactivating effect. The lower solubility of *n*-hexane, *n*-octane, and isooctane in water may account for their lower molecular toxicity to the enzyme.

The profiles of biocatalyst activity retention with incubation time were determined in both the single aqueous phase and the two-liquid phase systems without substrate. An exponential type of biocatalyst deactivation was observed in a neat aqueous buffer without organic solvent, with an estimated half-life of 21 h, indicating that some of the cells were lysed during the incubation period. When the cells were incubated in the biphasic system, the inactivation rate became greater and it could be approximately described by a first-order kinetic equation. There was no significant difference between the fitted first-order inactivation rate constants found for the enzyme incubated in the system with *n*-hexane (k_{in} = −3.57% activity per h), *n*-octane (k_{in} = −3.07% activity per h), or isooctane (k_{in} = −3.07% activity per h). Based on these results, isooctane was chosen as the best solvent because it exhibited the least negative effect on both enzyme activity and stability. Furthermore, it is a good solvent for GPE but a poor one for the diol, allowing easy recovery of the product.

The choice of a phase volume ratio for the two-phase bioconversion was made after examining the modulating effects of this parameter on the enzyme activity, stability, and enantiospecificity. The volume fraction of the organic phase in a two-liquid phase system may modulate the biohydrolysis of GPE in two ways. Firstly, different phase volume fractions may cause different interfacial areas due to altered degrees of emulsification under the same mixing condition, which may change the mass-transfer rate of the substrate to the aqueous phase as well as that of the product to the organic phase. Secondly, a change in the volume fraction may result in different organic phase effects on the microbial cells and therefore modulate the hydrolysis rate.

The production rates of the diol observed at various phase ratios were 7.8–12 nmol min^{-1} mg$_{DCW}$$^{-1}$, in which the specific product formation rate reached the maximum (12 nmol min^{-1} mg$_{DCW}$$^{-1}$) at a phase ratio ($\varphi_{o/w}$) of 1:5. A test of the cell hydrophobicity indicated a low degree of hydrophobicity of *B. megaterium* ECU1001 cells. Namely, the reaction catalyzed by the resting cells happens in aqueous bulk phase, not on the organic–aqueous interface. Therefore, the mass-transfer rate of epoxide from the organic phase to the aqueous phase may also play a great role in modulating the reaction rate. No significant difference in the residual activity of the cells was observed at the phase volume ratios tested, suggesting that phase volume ratio is not a sensitive parameter for modulating the enzyme stability in the biphasic system.

Carefully modulated enantioselectivity is perhaps the most attractive feature of enzymatic synthesis. However, the modulative effect of phase volume ratio on the enantioselectivity was rarely reported. From the results listed in Table 2.6, the enzymatic reaction in the single aqueous phase of potassium phosphate buffer (KPB, 50 mM, pH 8.0) containing 5% (v/v) of DMSO exhibited a relatively higher conversion but the lowest enantioselectivity as compared with those in aqueous–organic biphasic system. In addition, a decrease in the isooctane content of the two-phase system resulted in an obvious increase in the substrate conversion, while the highest enantioselectivity of the enzyme was observed at an isooctane–water volume ratio ($\varphi_{o/w}$) of 1:5. It is very interesting that the enantiomeric ratio (*E*-value) of the enzymatic reaction was significantly regulated up to 94.0 in the two-phase system, 2.4 times higher than that in the single aqueous phase system.

The stirred-tank reactor (STR) is one of the simplest and most widely used bioreactor types. For a preparative resolution of GPE, a mechanically stirred reactor was operated in batch mode under the above optimal conditions

Table 2.6   Modulating effect of the phase volume ratio on the enantioselectivity of BmEH.[a]

Phase ratio (organic–aqueous, v/v)	ee$_s$ (%)	ee$_p$ (%)	Conv. (%)	E^b
9:1	27.4	99.9	22.5	45.2
5:1	52.4	96.5	36.2	42.4
2:1	77.4	94.0	46.2	45.2
1:1	96.3	89.4	51.8	72.8
1:3	97.9	87.8	52.5	74.5
1:5	100.0	85.2	53.9	94.0
1:9	100.0	79.5	55.7	63.5
0:10^c	100.0	77.4	59.5	39.5

[a]Reaction conditions: epoxide, 150 μmol; cells, 3.85 mg (DCW); total volume, 12 mL; 30 °C, 180 rpm, 24 h.
[b]$E = \ln[(1 - c)(1 - ee_s)]/\ln[(1 - c)(1 + ee_s)]$.
[c]The reaction was carried out in a single aqueous phase (50 mM KPB, pH 8.0) containing 5% (v/v) DMSO.

(*i.e.*, 600 mM GPE dissolved in 20 mL of isooctane was mixed with 1560 mg of cells suspended in 100 mL of buffer). Figure 2.6 shows the time-dependent changes in the concentration and ee value of epoxide and diol in the reactor. The curve is typical of a highly enantioselective hydrolysis of a racemic substrate. The ee_s of the remaining epoxide increased from 0 to 100% after 24 h. In contrast, the ee_p of the diol produced decreased slightly from 87.4 to 80.3%. After 24 h, the concentration of residual epoxide was 44.5 mM (44.5% yield, 100% ee) and the concentration of diol was 55.5 mM (55.5% yield, 80.3% ee). The average productivity of the GPE in the enzyme reactor in this organic–aqueous system reached 48.9 mg GPE per h per $g_{biocatalyst}$,[32] which is 2.6-fold higher than that obtained in the single aqueous phase (18.8 mg GPE per h per $g_{biocatalyst}$).[29] Therefore, the use of an organic–aqueous two-phase system enabled a higher substrate (epoxide) load to reach a higher productivity. This method may be further developed into a practical technology for the production of optically pure (*S*)-GPE and other chiral epoxides.

For the first time, two novel epoxide hydrolases were discovered in *Vigna radiata* (mung bean) in our laboratory,[33] either of which can catalyze enantioconvergent hydrolysis of styrene epoxides (Scheme 2.3). Their regioselectivity coefficients are more than 90% for *p*-nitrostyrene oxide. Furthermore, the

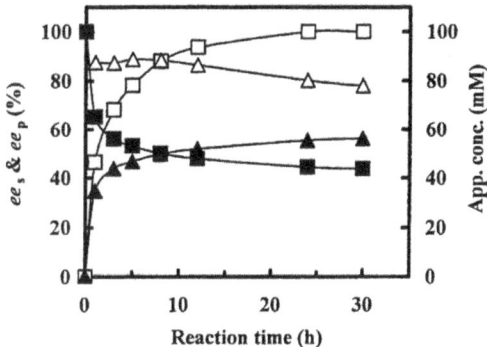

Figure 2.6 Biocatalytic resolution of glycidyl phenyl ether (GPE) with resting cells of *Bacillus megaterium* ECU1001 in a stirred reactor. Concentrations of the epoxide (■) and the diol (▲), and enantiomeric excesses of the epoxide (□) and the diol (△) were plotted *versus* the reaction time. Reaction conditions: epoxide, 12 mmol; cells, 1.56 g (DCW); isooctane, 20 mL; buffer, 50 mM KPB, pH 8.0, 100 mL; 30 °C, 600 rpm.

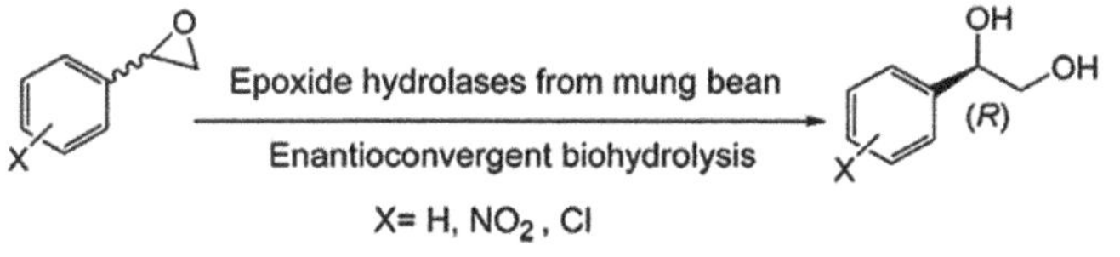

Epoxide hydrolases from mung bean
———————————————————
Enantioconvergent biohydrolysis

X= H, NO₂, Cl

Scheme 2.3 Enantioconvergent hydrolysis of styrene epoxides by *Vigna radiata* epoxide hydrolases.

crude mung bean powder was also shown to be a cheap and practical biocatalyst, allowing a one-step asymmetric synthesis of chiral (R)-diols from racemic epoxides, with up to >99% ee and 68.7% overall yield after recrystallization. The mechanism of enantioconvergent catalysis is illustrated in Figure 2.7.

Recently, one of the encoding genes (*vreh1*) was successfully cloned from the cDNA of *V. radiata* by RT-PCR and rapid amplification of cDNA ends (RACE). The *vreh1* gene constructed in a pET-28a(+) vector was then heterologously overexpressed in *Escherichia coli* BL21(DE3), and the recombinant protein was purified to homogeneity by nickel affinity chromatography.[34] It was shown that the epoxide hydrolase VrEH1 has an optimum activity at 45 °C and is very thermostable with a high inactivation energy (468 kJ mol^{-1}). Note that the activity of VrEH1 could be positively modulated up to ~340 or ~150% by adding 0.1% Triton X-100 or 0.1% Tween-20, respectively. VrEH1 shows an unusual ability of enantioconvergent catalysis for the hydrolysis of racemic *p*NSO, producing (R)-*p*-nitrophenyl glycol. It displays opposite regioselectivity toward (S)-*p*NSO (83% toward C_α), in contrast to (R)-*p*NSO (87% toward C_β). The K_M and k_{cat} of VrEH1 were 1.4 mM and 0.42 s^{-1} for (R)-*p*NSO, and 5.5 mM and 6.2 s^{-1} for (S)-*p*NSO.

A comparative study was made of asymmetric hydrolysis of styrene epoxide to (R)-1-phenyl-1,2-ethanediol by mung bean epoxide hydrolase in a biphasic system (*n*-hexane–buffer) containing hydrophilic ionic liquids (ILs). Compared to the biphasic system alone, the introduction of a small amount of hydrophilic ILs decreased the degree of non-enzymatic hydrolysis and increased the reaction rate by 22%. The ILs with a cation containing an alkanol group,

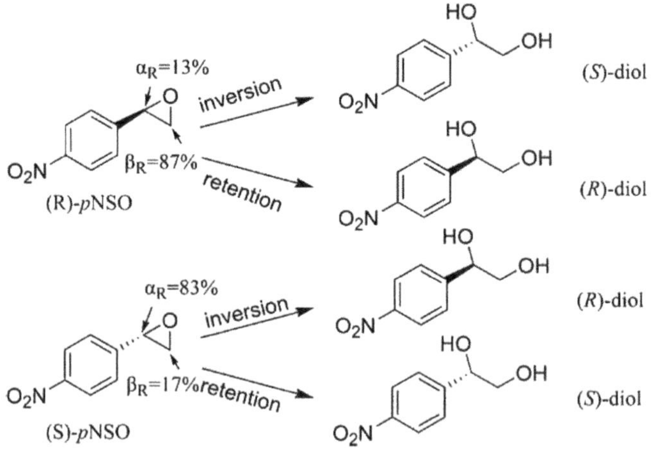

Figure 2.7   Regioselectivity of VrEH1 in enantioconvergent hydrolysis of (R)- and (S)-*p*NSO.[34] For the (R)-*p*NSO enantiomer, VrEH1 mainly attacks the C_α position of the epoxide, resulting in retention of the configuration and formation of (R)-diol with 87% probability, while it attacks the C_α position affording (S)-diol with merely 13% probability. On the other hand, for (S)-*p*NSO, VrEH1 mainly attacks C_α (83%), resulting in inversion of the configuration and formation of (R)-diol, whereas it attacks C_α affording (S)-diol with 17% probability.

including [C$_2$OHMIM][BF$_4$] and [C$_2$OHMIM][T$_f$O], and the choline amino acid ILs such as [Ch][Arg] and [Ch][Pro], were suitable modulators for the reaction due to their good biocompatibility with the enzyme, leading to a high initial rate (0.99–1.25 µmol min^{-1}) and high product ee$_s$ (~95%). When the substrate loading was 30 mM, the product enantiopurity was improved up to ≥95% ee in the presence of ILs, compared with 90% ee without ILs.[35]

2.3 Chemical Modulators for Promoting Glycosidase Reactions

2.3.1 Modulating Effect of Non-ionic Surfactants on Ginsenoside Biohydrolysis

Ginseng is one of the most well-known and widely used Chinese traditional medicines. Ginseng saponins or ginsenosides have been reported to be one of the most important physiologically active ginseng component with a variety of biological effects. Among them, ginsenside Rh2 is becoming increasingly attractive recently due to its suppressing effect on growth of various cancer cells. However, the contents of naturally occurring ginsenoside Rh2 in red ginseng and wild ginseng are only 10 and 30 ppm, respectively. Compared to the structure of other protopanaxadiol-type saponins (*e.g.*, Rg3) with higher contents in the ginseng, ginsenoside Rh2 has the same aglycone, but a different sugar moiety. Therefore, it would be a feasible method to produce ginsenoside Rh2 by modifying the sugar moiety of those ginsenosides that are affluent or easily available in ginseng. The selective hydrolysis of ginsenoside Rg3 using a specific β-glucosidase is considered to be promising for practical application (Scheme 2.4).

A novel β-glucosidase (FPG) from *Fusarium proliferatum* ECU2042 was successfully purified to homogeneity with a 506-fold increase in specific activity. Its molecular mass was estimated to be approximately 78.7 kDa, with two homogeneous subunits of 39.1 kDa, and its p*I* (isoelectric point) was 4.4 as indicated by two-dimensional electrophoresis. The optimal activities of FPG occurred at pH 5.0 and 50 °C. The enzyme was stable at pH 4.0–6.5 and temperatures below 60 °C, and the inactivation energy (E_{in}) for FPG was 88.6 kJ mol^{-1}.

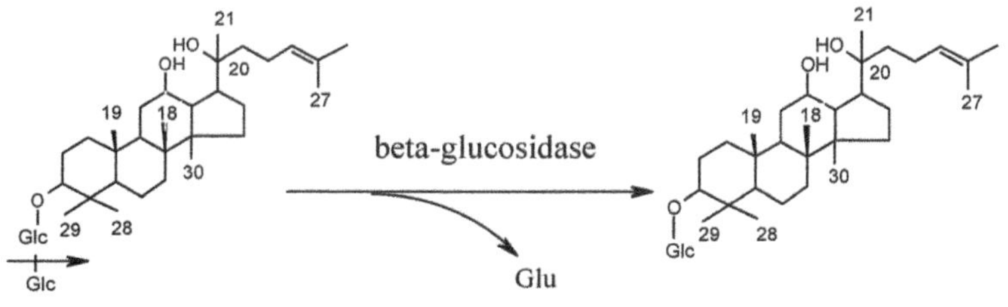

Scheme 2.4 Enzymatic transformation of ginsenside Rg3 to Rh2.

Table 2.7   Symbols and main components of selected nonionic surfactants.

Surfactant	Main component
Synperonic PE/F68	Poly(ethylene glycol)-*block*-poly(propylene glycol)-*block*-poly(ethylene glycol)
Synperonic PE/L61	Poly(ethylene glycol)-*block*-poly(propylene glycol)-*block*-poly(ethylene glycol)
Synperonic NP 10	Polyethylene glycol nonylphenyl ether
PEGDME 250	Polyethylene glycol 250 dimethyl ether
PEGMME 350	Polyethylene glycol 350 monomethyl ether
PEGDME 400	Polyethylene glycol 400 dimethyl ether
Triton X-45	Polyethylene glycol 4-*tert*-octylphenyl ether
Trion X-100	Polyoxyethylene (10) octylphenyl ether
Tween-80	Polyoxyethylene (80) sorbitan monooleate
IGEPAL CA-630	Octylphenoxypolyethoxyethanol
OP-10	Alkylphenol ethoxylates
Brij 92V	Diethylene glycol oleyl ether

Interestingly, the purified enzyme exhibited a very low activity towards *p*-nitrophenyl β-D-glucoside (*p*NPG), and almost no activity towards cellobiose, however, relatively high activity was observed on ginsenoside Rg3. The enzyme hydrolyzed the 3-C,β-(1→2)-glucoside of ginsenoside Rg3 to produce ginsenoside Rh2, but did not sequentially hydrolyze the β-D-glucosidic bond of Rh2. The K_m and V_{max} values of FPG for ginsenoside Rg3 were 2.37 mM and 0.568 μmol per h per mg protein, respectively. In addition, this enzyme also exhibited significant activities toward various alkyl glucosides, aryl glucosides, and several natural glycosides.[36]

Several nonionic surfactants (Table 2.7) were tried to improve the enzymatic hydrolysis of ginsenoside Rg3 into Rh2 which was catalyzed at 50 °C and pH 5.0 by a crude glucosidase extracted from *Fusarium* sp. ECU2042. Among the biocompatible nonionic surfactants, polyethylene glycol 350 monomethyl ether (PEGMME 350) showed the best reaction performance. After optimizing some factors in the reaction, the conversion of Rg3 (5 g L⁻¹) with 10 g L⁻¹ crude enzyme reached almost 100% in the presence of the nonionic surfactant (7.5%, w/v), which was 25% higher than that in buffer without any modulator (Figure 2.8). Furthermore, the enzyme stability was seldom impaired by the non-ionic surfactant.[37]

2.3.2   Enzymatic Synthesis of Alkyl Glycosides by Reverse Hydrolysis

A facile method for the enzymatic glycosylation of 4-substituted benzyl alcohols and tyrosol with glucose in a monophasic aqueous–dioxane medium was reported,[38] using a crude meal of apple seed as a new catalyst. The corresponding β-D-glucosides (**1–7**) were synthesized in moderate yields (13.1–23.1%), among which the naturally precious salidroside (**7**) was easily obtained in 15.8% yield (Table 2.8).

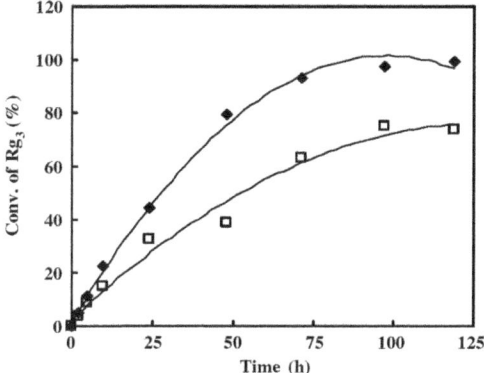

Figure 2.8 Progress curves of ginsenoside Rg3 hydrolysis by *Fusarium* sp. ECU2042 at 50 °C. The reaction mixture was composed of 0.2 mL of NaAc–HAc buffer (pH 5.0), 2 mg of lyophilized crude enzyme from *Fusarium* sp. ECU2042 and 1 mg of ginsenoside Rg3. Filled diamonds, with surfactant (PEGMME 350, 7.5%, w/v); open squares, without surfactant.

Table 2.8 Enzymatic synthesis of β-glucopyranosides from glucose (0.25 M) and various alcohols using apple seed meal in buffer–dioxane (1 : 10, v/v).

1 n=1, X=H
2 n=1, X=OCH3
3 n=1, X=CH3
4 n=1, X=NO2
5 n=1, X=F
6 n=1, X=Cl
7 n=2, X=OH

Entry	n	X	Alcohol state	Yield[a] (%)
1	1	–H	Liquid	20.4
2	1	–CH$_3$O	Liquid	23.1
3	1	–CH$_3$	Solid	16.7
4	1	–NO$_2$	Solid	13.1
5	1	–F	Liquid	18.1
6	1	–Cl	Solid	16.1
7	2	–OH	Solid	15.8

[a]Isolated yield.

A monophasic organic–water system for the efficient enzymatic synthesis of β-D-glucopyranoside by reverse hydrolysis was designed and optimized. *p*-Nitrobenzyl alcohol (*p*NBA), selected as a model substrate alcohol, was readily glucosylated with D-glucose through reverse hydrolysis using almond β-D-glucosidase in a monophasic aqueous–organic medium, producing a new glucoside, *p*-nitrobenzyl β-D-glucopyranoside (*p*NBG). The formulation of different buffers, organic solvents, and water contents was adjusted. Buffer type and pH modulated the initial reaction rate but had little effect on the final yields. The ratio of organic solvent to water plays a crucial role in regulating the reaction equilibrium toward synthesis, but a minimum amount of water is necessary to rescue the enzyme activity (Figure 2.9). Dioxane, which was previously known as an unsuitable solvent for β-D-glucosidase-catalyzed reactions, was found to be the most appropriate medium for this synthetic procedure. The reaction equilibrium and enzyme stability in the reaction medium were also investigated. Under the optimal reaction conditions, *i.e.* 90% dioxane (v/v) + 10% buffer (Na$_2$HPO$_4$–KH$_2$PO$_4$, 70 mM, pH 6.0) with an alcohol-to-glucose molar ratio of 9 : 1, *p*-nitrobenzyl β-D-glucopyranoside was produced with a maximum yield (13.3%).[39]

Salidroside is a natural glycoside with pharmacological activities of resisting anoxia, microwave radiation, and fatigue, alleviating lack of oxygen, and postponing aging. In this work, salidroside and other natural glucosides such as cinnamyl *O*-β-D-glucopyranoside and 4-methoxybenzyl *O*-β-D-glucopyranoside were efficiently synthesized *via* an environmentally benign and energy economical process (Scheme 2.5). In the synthetic process, apple seed, easily available from fruit processing factory waste, was employed as a natural and green catalyst. Moreover, all of the catalyst, solvent and excessive substrate was reused or recycled. The biocatalytic reaction was carried out in a clean and less toxic medium of aqueous *tert*-butanol and the glucoside produced was selectively removed from the reaction mixture by alumina column adsorption, making excessive substrate (aglycon) recyclable for repeated use in the next batch of reactions. To improve the biocatalyst stability, apple seed meal was further cross-linked by glutaraldehyde, yielding a net-like porous structure within which the dissociating proteins were immobilized, resulting in improved permeability of the biocatalyst. After the simple cross-linking treatment, the half-life of the apple seed catalyst was significantly prolonged from 29 days to 51 days. The productivity of the bioreactor in the case of salidroside reached *ca.* 1.9 g per L per day, with the product purity up to 99.3% after post-reaction refinement.[40]

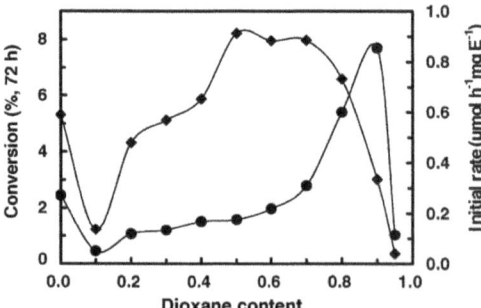

Figure 2.9   Enzymatic synthesis of *p*-nitrobenzyl-D-glucopyranoside (pNBG) by reverse hydrolysis in a monophasic dioxane–water system. The reaction was performed in 1.0 mL of dioxane–buffer medium (Na$_2$HPO$_4$–KH$_2$PO$_4$, 70 mM, pH 6.0) by shaking a mixture of 0.25 mmol glucose, 1.0 mmol pNBA and 5 mg of enzyme powder at 50 °C and 160 rpm. Symbols: (♦) initial rate; (●) final conversion.

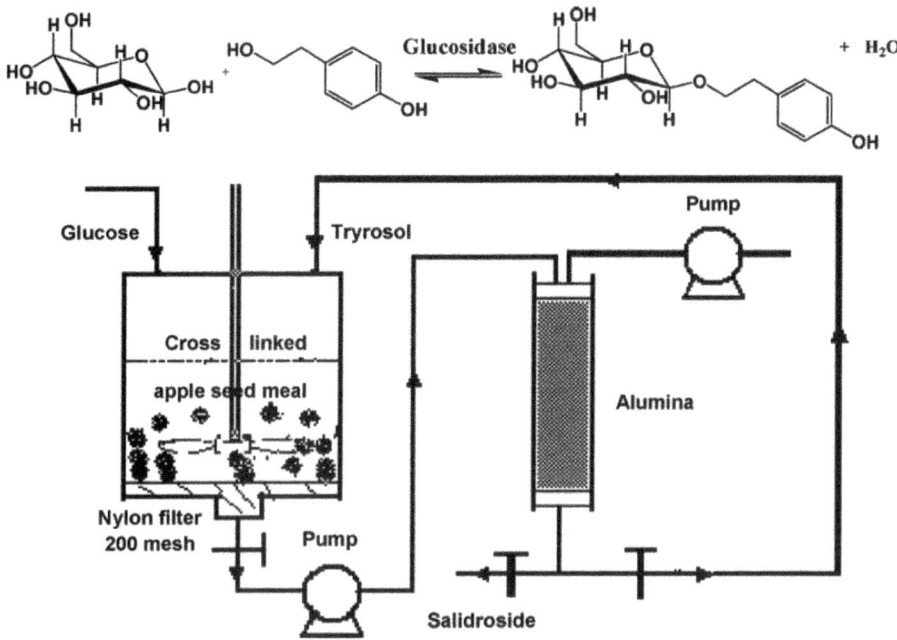

Scheme 2.5 Enzymatic synthesis of salidroside catalyzed by cross-linked apple seed meal integrated with alumina adsorption.

Glycoconjugates can be artificially synthesized by combinatorial biocatalysis. An example given herein describes the construction of a glycoconjugate array by using glycosidase and lipase in non-aqueous media (Scheme 2.6). This array was started from glucose, with three aryl alcohols as the aglycone moiety of glycosides and five acids or esters as acyl donors for combinatorial acylation of glycosides, affording a three-dimensional array containing about 30 members with diverse structures. The array would be more abundant if more aglycones and acyl donors with other structures were adopted. Indeed, diverse classes of carbohydrates besides glucose can also be employed for generating diverse glycoconjugates due to their different roles in numerous physiological responses. The composition and distribution of the demonstrative glycoconjugate array were detected and evaluated by HPLC-MS with electrospray ionization, and the distribution of the artificial array can be adjusted by changing the molar ratio of the auxiliary materials.[41]

2.3.3 Enzymatic Synthesis of Alkyl Glycosides with Ionic Liquid as a Modulator

Enzymatic synthesis of various arylalkyl β-D-glucopyranosides catalyzed by prune (*Prunus domestica*) seed meal *via* reverse hydrolysis in the mixture of organic solvent, ionic liquid (IL) and phosphate buffer was described (Scheme 2.7). Among four hydrophilic organic solvents tested, ethylene

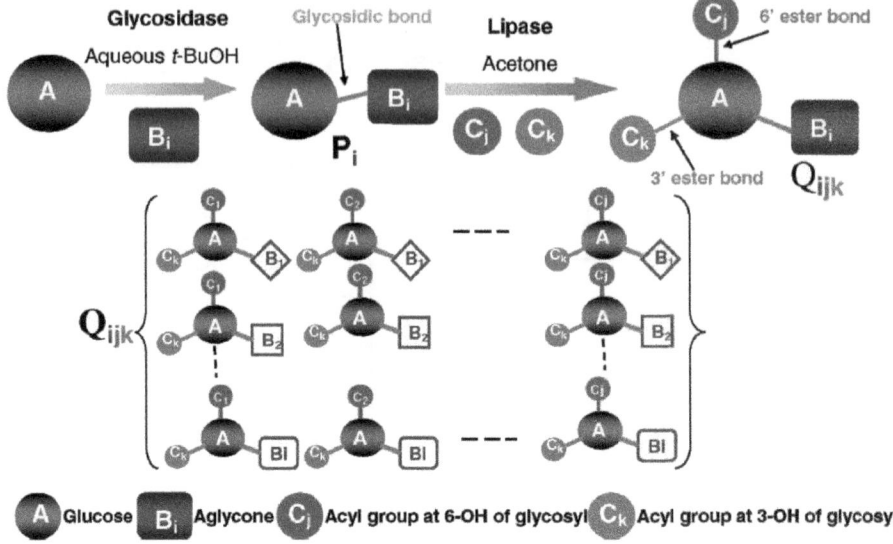

Scheme 2.6   Combinatorial biocatalysis of glycosidase and lipase in nonaqueous media for constructing a three-dimensional array of glycoconjugates (Q_{ijk}).

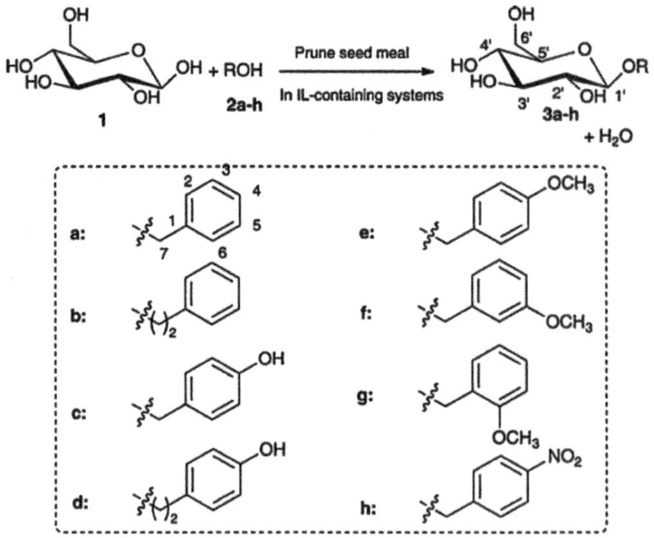

Scheme 2.7   Enzymatic synthesis of various arylalkyl β-D-glucopyranosides in IL-containing systems.

glycol diacetate (EGDA) was found to be the most suitable for enzymatic synthesis of salidroside, a bioactive compound of commercial interest, from D-glucose and tyrosol. The modulating effects of the nature and content of ionic liquids on the enzymatic glucosylation were studied (Table 2.9). The addition of a suitable amount of ILs including denaturing ones

Table 2.9 Modulating effect of ILs on enzymatic synthesis of salidroside.[a]

Entry	IL	V_0 (mM h^{-1})	Yield (%)
1	None	3.5	18
2	[BMIm]BF$_4$	3.0	21
3	[BMMIm]BF$_4$	3.0	21
4	[HMIm]BF$_4$	3.3	20
5	[BMIm]PF$_6$	3.2	17
6	[BMIm]C$_8$SO$_4$	2.9	20
7	[MMIm]MeSO$_4$	3.0	21
8	[BMIm]I	3.1	22
9	[BMIm]Cl	0	0
10	[EOEMIm]Cl	0.5	9
11	[AcOEMIm]Cl	0.4	2

[a]Conditions: 0.5 mmol glucose, 5 mmol tyrosol, 5.5 U prune seed meal, EGDA: 10% (v/v), IL: 10% (v/v) phosphate buffer (pH 6.0, 50 mM), total volume 2 mL, 50 °C, 200 rpm. Reaction time: 72 h.

was favorable to modulate the reaction equilibrium toward the direction of synthesis, thus improving the yields. Among the examined ILs, the novel IL [BMIm]I proved to be the best. This IL was applied as the modulator in biocatalysis for the first time. The yields were enhanced by 0.2-fold to 0.5-fold after the addition of 10% (v/v) [BMIm]I. In the 10% (v/v) [BMIm] I-containing system, the desired arylalkyl β-D-glucopyranosides were synthesized with 15–28% yields, among which salidroside was obtained with 22% yield.[42]

2.4 Modulating Enzymatic Reaction by Immobilization Techniques

An immobilized enzyme is an enzyme that is attached to an inert, insoluble material to provide increased resistance to changes in reaction conditions such as pH or temperature. Additionally, it allows enzymes to be trapped in a reactor throughout the reaction, following which they are easily separated from the products and may be recycled for utilization. Therefore, immobilized biocatalysts may represent the earliest and easiest ways to modulate the performance of native enzymes for application in industry due to its economical end.

There are three different ways to immobilize an enzyme, which are as follows:

  i) Adsorption: enzyme is attached to the outside of an inert material.
 ii) Entrapment: the enzyme is trapped in insoluble beads or microspheres, such as calcium alginate beads.
iii) Cross-linking: the enzyme is covalently bonded to a matrix through a chemical reaction. Enzymes may also be immobilized by cross-linking with themselves or binding to a surface with non-covalent or covalent protein tags.

2.4.1   Reactors with Immobilized Enzymes on Classic Supports

Various materials have been used as supports for the immobilization of a *Candida rugosa* lipase (Lipase OF).[43] The enzyme adsorbed on silica gel showed the maximum activity recovery (37%), and was thus used to construct a packed bed reactor for the continuous production of (S)-ketoprofen. The performance of the bioreactor containing 4.0 g of immobilized biocatalyst was evaluated as a function of parameters such as flow rate, the ratio of height to diameter (*H/D*) and substrate concentration. The space velocity (SV) of the reactor showed a linear increase when the flow rate was increased in the range of 5.9–25 mL h^{-1} (SV = 0.9–3.9 h^{-1}) and leveled out at values for flow rate between 25 (SV = 3.9 h^{-1}) and 64 mL h^{-1} (SV = 10.2 h^{-1}). The initial rate remained the same when the *H/D* was varied from 2.9 to 38.6. The initial rate rose linearly as the substrate concentration was increased from 5 to 20 mg mL^{-1} and remained almost at the same level when the substrate concentration was between 20 and 50 mg mL^{-1}. Using a packed bed bioreactor, optically pure (S)-ketoprofen (>99% ee) was produced with a conversion of 30% and productivity of 1.5 mg per g$_{biocatalyst}$ per h.

The lipase from *Serratia marcescens* ECU1010 (*Sm*L) was capable of enantioselectively catalyzing the synthesis of many chiral drug precursors. The immobilization of *Sm*L on appropriate supporting materials and its performance in bioreactor were investigated.[44] Chitosan, Celite 545, and DEAE-cellulose were found to be the ideal supports among 8 carriers tested with respect to enzyme load and activity recovery of lipase. When *Sm*L was immobilized, significant improvements of stability against pH, thermal, and operational deactivation were observed with all the 3 better supports, and the best stability was observed when the lipase was immobilized on glutaraldehyde activated chitosan. As for the effect of organic solvent on the biphasic reaction system, the hydrolytic activity of the immobilized lipase on *trans*-3-(4′-methoxyphenyl)glycidic acid methyl ester ((±)-MPGM) observed in isopropyl ether was 6 and 3 times higher than those in toluene and methyl *tert*-butyl ether. The lipase catalyzed production of (−)-MPGM by enzymatic resolution of (±)-MPGM with chitosan–*Sm*L in an isopropyl ether–water biphasic system was carried out in a 2 L stirred-tank reactor (Figure 2.10). Batch operation was a more efficient operation mode for the enantioselective hydrolysis of (±)-MPGM, producing enantiopure (−)-MPGM in a 44.3% overall yield, in contrast to 29.3% in a continuous reactor.

Optically active (R)-(+)-α-hydroxy-γ-butyrolactone (R-HBL) was produced *via* enantioselective hydrolysis of racemic HBL using a lactonase extracted from *Fusarium proliferatum* ECU2002 (FpL).[45] Different carriers were examined for immobilizing FpL and the highest activity was observed when the enzyme was adsorbed onto cotton cloth followed by cross-linking with glutaraldehyde. A fibrous bed reactor (FBR) (Scheme 2.8) was constructed by packing a piece of cotton cloth (*ca.* 2 g) coiled together with a wire net into

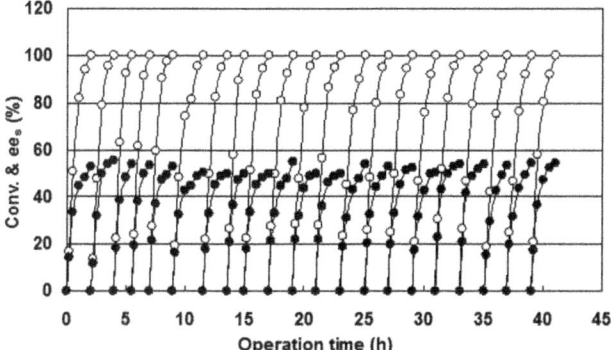

Figure 2.10 Repeated production of (−)-MPGM in a 2 L batch stirred tank reactor. (●) Conversion; (○) ee$_s$ (ee of substrate). Reaction was performed in a 2 L bioreactor at 30 °C and 200 rpm using an isopropyl ether–water biphasic system with 100 mM (±)-MPGM and 900 U of chitosan–SmL. The pH was automatically controlled at 8.0–8.3 by titrating 2% ammonia. The immobilized lipase was reused in new reactions after ee$_s$ of the last run reached 99.9%. From the 7th batch of reaction, 15 g of fresh immobilized lipase (corresponding to 10% of the initial amount, *ca.* 90 U) was supplemented at the beginning of each reaction cycle. After the 12th batch of reaction, 15 g of used immobilized lipase was substituted by 15 g of fresh lipase (*ca.* 90 U). The partial renewal of the immobilized biocatalyst was continued for 10 cycles (12th–21st).

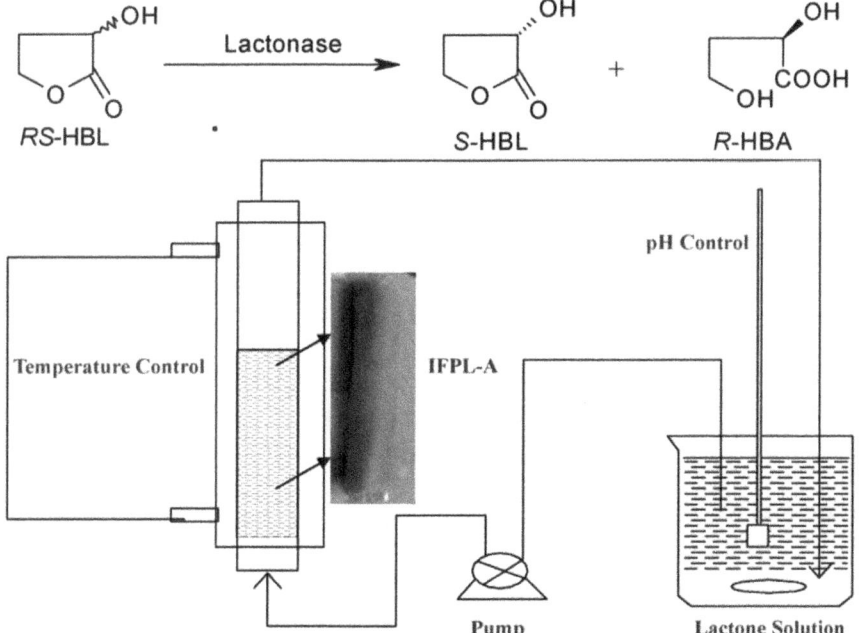

Scheme 2.8 Reaction device employed for enantioselective hydrolysis of racemic α-hydroxy-γ-butyrolactone by immobilized lactonase from *Fusarium proliferatum* ECU2002.

a glass column (ϕ 0.5 cm × 12 cm) thermostated at 30 °C. Kinetic resolution of *RS*-HBL was carried out semi-continuously in the FBR by circulating a racemic lactone solution through the reactor at a certain flow rate. The performance and productivity of the FBR were evaluated by adjusting several critical parameters, including enzyme load and initial *RS*-HBL concentration. Immobilized FpL (IFPL, *ca.* 40 U) per 50 mL of working volume turned out to be the optimal enzyme load, and the most suitable substrate concentration was 750 mM at 30 °C with an appropriate height to diameter (*H*/*D*) ratio (5.0). The IFPL-catalyzed kinetic resolution of *RS*-HBL was successfully operated in the FBR for 60 batches, with an average productivity of 2.48 g L^{-1} h^{-1} (*R*-HBL) in high optical purity (90.0–96.4% ee) in the case of semi-continuous operation.

2.4.2   Performance of Nanoparticle-Supported Enzymes

Magnetic Fe$_3$O$_4$ nanoparticles were prepared by chemical coprecipitation method and subsequently coated with 3-aminopropyltriethoxysilane *via* silanization reaction (Scheme 2.9). The synthesized materials were characterized by transmission electron microscopy (TEM) and Fourier transform infrared spectroscopy (FTIR). With glutaraldehyde as the coupling agent, the lipase from *Serratia marcescens* ECU1010 (*Sm*L) was successfully immobilized onto the amino-functionalized magnetic nanoparticles.[46] It was shown that the immobilized protein load could reach as high as 35.2 mg protein per g support and the activity recovery was up to 62.0%. The immobilized lipase demonstrated a high enantioselectivity toward (+)-MPGM (with an *E*-value of 122) and it also displayed improved thermal stability as compared to the free lipase. When the immobilized lipase was employed to enantioselectively hydrolyze (±)-*trans*-3-(4-methoxyphenyl) glycidic acid methyl ester [(±)-MPGM] in a water–toluene biphasic reaction system for 11 consecutive cycles (105 h in total), the majority (59.6%) of its initial activity was retained, indicating excellent stability in practical operation.

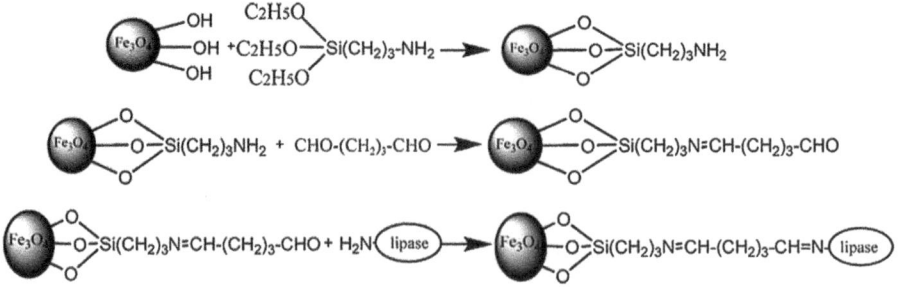

Scheme 2.9   The strategy used to immobilize lipase onto Fe$_3$O$_4$ nanoparticles.

Glycolate oxidase was isolated from *Medicago falcata* Linn. after screening 13 kinds of C3 plant leaves, with higher specific activity than the enzyme from spinach. The *M. falcata* glycolate oxidase (*MfGO*) was partially purified and then immobilized onto hydrothermally synthesized magnetic nanoparticles *via* physical adsorption.[47] The maximum load of MfGO was 56 mg g^{-1} support and the activity recovery was 45%. Immobilization of *MfGO* onto magnetic nanoparticles enhanced the enzyme stability, and the optimum temperature was significantly increased from 15 °C to 30 °C. The immobilized biocatalyst was successfully used in a batch reactor for repeated oxidization of glycolic acid to synthesize glyoxylic acid, retaining *ca.* 70% of its initial activity after 4 cycles of reaction at 30 °C for nearly 70 h, and its half-life was calculated to be 117 h (Figure 2.11).

Uridine diphosphate glucose (UDP-Glc) serves as a glucosyl donor in many enzymatic glycosylation processes. A multiple enzyme, one-pot, biocatalytic system was developed for the synthesis of UDP-Glc from low cost raw materials: maltodextrin and uridine triphosphate. Three enzymes needed for the synthesis of UDP-Glc (maltodextrin phosphorylase, glucose-1-phosphate thymidylyltransferase, and pyrophosphatase) were expressed in *Escherichia coli* and then immobilized individually on amino-functionalized magnetic nanoparticles. The conditions for biocatalysis were optimized and the immobilized multiple-enzyme biocatalyst could be easily recovered and reused up to five times in repeated syntheses of UDP-Glc. After a simple purification, approximately 630 mg of crystallized UDP-Glc was obtained from 1 L of reaction mixture, with a moderate yield of around 50% (UTP conversion) at very low cost.[48]

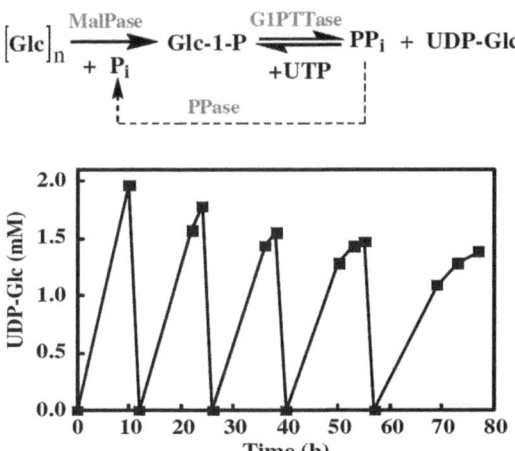

Figure 2.11 Time course curves of the repeated UDP-Glc production using three enzymes (MalPase, GlPTTase, and PPase) immobilized on amino-functionalized magnetic nanoparticles. Reaction conditions: MalPase, 18 U L^{-1}; GlPTTase, 100 U L^{-1}; PPase, 25 U L^{-1}; maltodextrin, 5% (m/v); UTP, 2 mM; MgCl$_2$, 10 mM; buffer, sodium phosphate, pH 7.5, 100 mM; 30 °C; total volume, 200 mL.

2.4.3 Reactions Using Crosslinked Enzyme Aggregates

Cross-linking of enzyme aggregates is a promising method for enzyme immobilization and modulation. A recombinant esterase (EC 3.1.1.1) cloned from
Bacillus subtilis 0554 (BsE) was carrier-freely immobilized with cross-linked
enzyme aggregates. The conditions for preparing the cross-linked aggregates
of BSE (CLA-BsE) were optimized, including the type and concentration of
precipitants, and the concentration of cross-linker, and a simple and efficient procedure for preparing CLA-BsE (Figure 2.12) was developed, consisting of a precipitation step with 0.5 g mL^{-1} (NH$_4$)$_2$SO$_4$ and a cross-linking step
with 60 mM glutaraldehyde for 3 h. The resultant CLA-BsE recovered about
70% of the initial free BsE activity. The thermal stabilities of the immobilized enzyme at 30 °C and 50 °C were >360 and 14 times higher than those
of free BsE, respectively. More importantly, the operational stability of CLA-
BsE was also considerably improved. In the kinetic resolution of DL-menthyl acetate to produce L-menthol with CLA-BsE, >94% ee$_p$ was achieved at

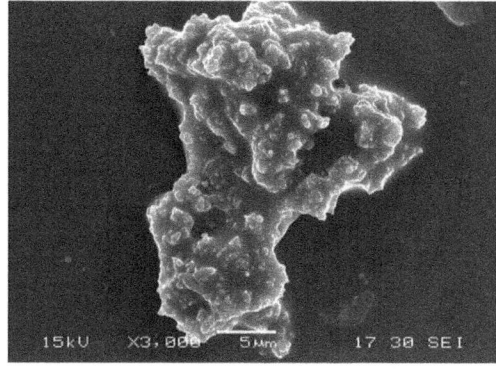

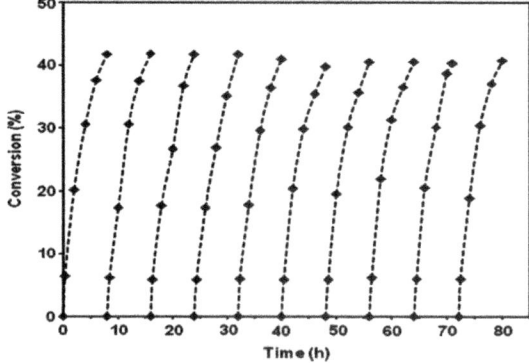

Figure 2.12   SEM photograph and reusability of CLA-BsE in the bioresolution of
DL-menthyl acetate. Reaction was performed at 30 °C in aqueous solution containing 10% EtOH and 1 M DL-menthyl acetate. The pH was
kept at 8.0. After each batch of reaction, the CLA-BsE was centrifuged
and washed after each use, and then resuspended again in an ensuing
fresh reaction mixture.

>40% conversion, and the CLA-BsE could be reused 10 times with only about 8% activity loss. Therefore, the new biocatalyst immobilized through the methodology of CLEAs could significantly reduce the manufacturing cost of L-menthol and would be more beneficial for its practical applications.[49]

In another work, cross-linked enzyme coaggregates of *Serratia marcescens* lipase with polyethyleneimine (CLECAs-*Sm*L–PEI) were prepared using polyethyleneimine (PEI) as the coprecipitant and glutaraldehyde as the crosslinking reagent. The crude lipase solution at a low protein concentration (0.1 mg mL^{-1}) with PEI at a mass ratio of 3:1 (PEI/protein, w/w) was found to be adequate for the coprecipitation of *Sm*L. After cross-linking of the coaggregate of *Sm*L–PEI with 0.2% (w/v) glutaraldehyde at ambient temperature, over 70% of the total lipase activity was recovered. Compared with the free *Sm*L, the optimum temperature of the CLECAs-*Sm*L–PEI was enhanced from 50 °C to 60 °C and its thermal stability was also significantly improved. CLECAs-*Sm*L–PEI showed excellent operational stability in repeated use in biphasic aqueous–toluene system (Figure 2.13) for the asymmetric hydrolysis of

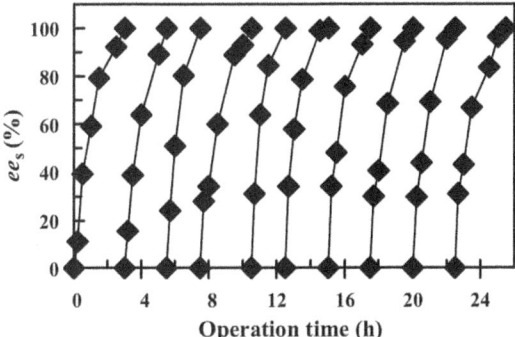

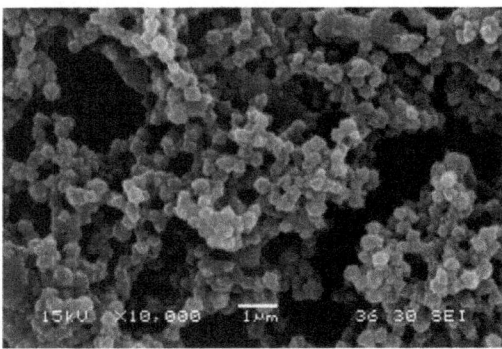

Figure 2.13 Repeated batch bioresolution (left picture) of (±)-MPGM using CLECAs-SmL–PEI (right picture, SEM image). The reaction was performed at 30 °C in an aqueous–toluene biphasic system with 0.1 M (±)-MPGM in toluene. After each batch of reaction, the aqueous phase containing CLECAs-SmL–PEI was withdrawn and mixed with fresh organic phase containing (±)-MPGM for the next batch of reaction.

trans-3-(4′-methoxyphenyl)glycidic acid methyl ester (MPGM), without significant inactivation after 10 rounds of repeated use.[50]

2.5 Modulating Enzyme Reaction by Amino Acid Substitutions

2.5.1 An Overview of Amino Acid Substitution Techniques

Traditionally, the enzyme sources are limited to extraction of natural products. The low production rate of enzymes leads to relatively high cost for a biotransformation process. With the development of recombinant DNA and DNA sequencing techniques, it is feasible to produce recombinant enzymes in various host expression systems, such as *E. coli*, *Bacillus subtilis*, yeasts or filamentous fungi.

Although these host systems are easy to cultivate and can produce satisfying amounts of functional enzymes, the intrinsic defect of enzymes still impedes their industrial application. An initially isolated enzyme, either from environmental samples or from genomic databases, has evolved naturally and been selected under the stress conditions among diverse populations of gene variants whose sequence variations rendered possible different enzyme properties. However, the fitness of a "wild-type" enzyme is usually not satisfactory, considering the activity, thermostability or selectivity of an enzyme for a specific chemical reaction.

All enzymes are proteins, which are linear sequences of amino acids linked by peptide bonds. The folding of these sequences determined the secondary structure (such as α-helix, β-sheet or β-turn) and tertiary structure. Therefore, the properties of an enzyme are actually presumed from its sequence of amino acids. Some amino acids, dubbed "hot spots", especially the ones in the active site where substrate binds, are sensitive to catalytic properties of an enzyme. Substitution of these important amino acids can significantly improve the activity or enantioselectivity toward a certain reaction. Protein stability is also maintained by the intramolecular and intermolecular interactions between residues of amino acids, including van der Waals forces, hydrophobic forces, electrostatic forces, hydrogen bonds and disulfide bonds. Detailed analysis of these amino acids, usually located in the protein surface, sheds light on the protein design for better thermostability.

An amino acid sequence is inherited from coding DNA with messenger RNA transferring the genetic information. An amino acid is faithfully encoded by a combination of three nucleotide bases according to the genetic code (codons). Since there are 4 kinds of nucleotide bases, the 20 kinds of basic amino acids are determined by 64 (4^3) possible codons, including 1 start codon (AUG) and 3 stop codons (UAA, UAG and UGA). Although chemically changing the residues of amino acids might be laborious and time-consuming, applied molecular biology techniques make such changes feasible by engineering the coding DNA sequence. To date, numerous techniques have been developed for amino acid substitutions, and peptide fragment insertion and deletion. The common methods are discussed below.

Error-prone PCR (epPCR) is perhaps one of the most widely used techniques for developing better and robust biocatalysts. It introduces mutations into the primary structure (the amino acid sequence) without knowing in advance the catalytic mechanism and three-dimensional structure of an enzyme. Generally, epPCR involves several repeated rounds of random mutagenesis, protein expression and screening until the desired property is achieved (Scheme 2.10). The basic concept of mutagenesis relies on modestly reducing the replication fidelity of polymerase in the polymerase chain reaction (PCR) process. The common methods for controlling the PCR fidelity comprise adjusting the concentration of $MnCl_2$ supplemented, using a low fidelity polymerase, or adding an unbalanced mixture of dNTPs. The main drawback of epPCR is that it must be coupled with a moderate to high throughput screening process, since each round usually generates 10^3 to 10^6 variants. However, many reactions are not suitable for prompt screening of variants. In addition, the two-amino-acids changes would theoretically generate more than 30 million variants in a 300-amino acid protein, which makes multi-site variants screening notoriously difficult for epPCR.[51]

DNA shuffling offers an alternative technique for directed evolution of biocatalysts. Compared with point mutagenesis using epPCR, DNA shuffling

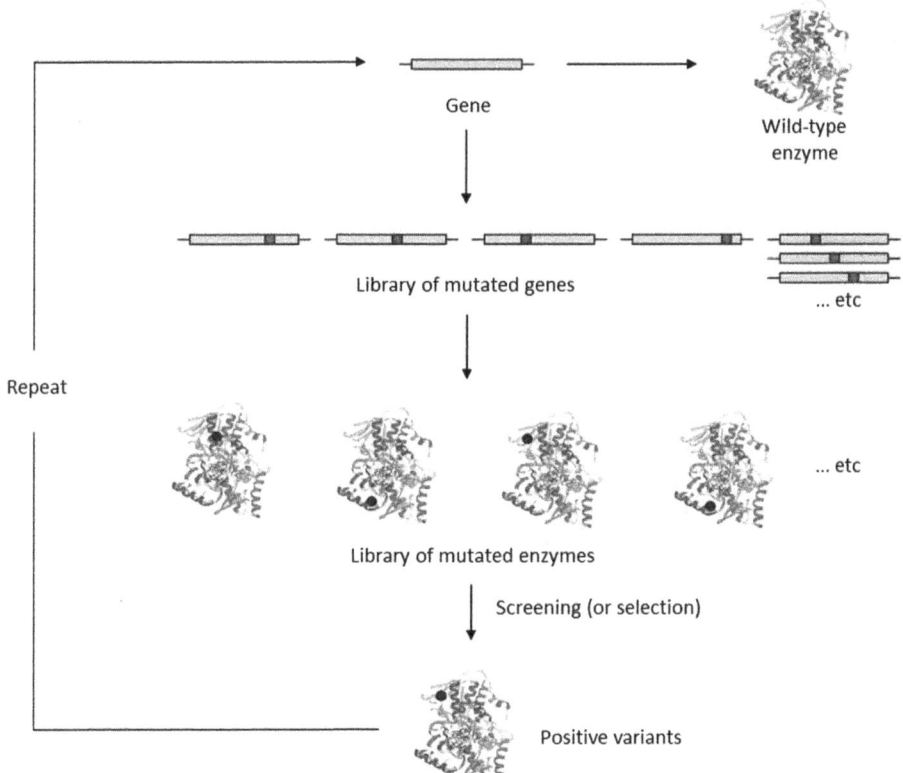

Gene

Wild-type enzyme

Library of mutated genes

... etc

Repeat

... etc

Library of mutated enzymes

Screening (or selection)

Positive variants

Scheme 2.10 Schematic outline of a typical error-prone PCR expepriment.[52]

takes advantage of gene diversity by recombining fragments of homologue gene sequences. The basic approach involves randomly cutting similar gene sequences into fragments and recombining these fragments into chimeric mutated genes with various sequence diversities (Scheme 2.11). The homologue genes can be naturally obtained from various species or mutated gene libraries. This method introduces more randomness into sequences compared with epPCR, which only introduces mutations stepwise.

Random mutagenesis requires screening a large amount of mutant transformants. Moderate to high throughput screening methods are developed for the routine and unbiased discovery of better variants, including agar plate screening, 96-well plate screening and fluorescence-activated cell sorting (FACS). These techniques usually involve color- or fluorescence-generating reactions (either from reaction products or reporter genes) for quantitative evaluation. Techniques such as promoter design,[54] chaperonin coexpresion,[55] or cell surface display[56] are employed to improve convenience, sensitivity, and efficiency of the screening process. Promoter design (alteration of promoter-coding nucleic acids), for instance, can be used to modulate gene transcription and therefore the amount of expressed protein. It circumvents the dilution of thousands of enzyme samples in the screening process, which will significantly save time and reduce operational error. In addition, promoter substitution can also be used for controlling reporter gene (*e.g.* green fluorescence protein) expression to improve the feasibility of the screening system.

Another approach is rational protein design and semi-rational design. Although the experimental design requires less efforts, directed evolution techniques, including epPCR, DNA shuffling and the combinatorial approaches, usually generate a large number of mutated genes, making it difficult to screen or select the superior variants. Rational enzyme design, on the other hand, focuses on spotting important amino acid residues, dubbed "hot spots", in target proteins. Theoretically, rational mutagenesis on the key amino acid spots generates smarter variant libraries, which therefore require less screening efforts. However, rational design of a "bespoke" enzyme is sometimes hampered by scarce knowledge on a protein's tertiary structure or catalytic mechanism, and the complexity of the protein itself. Recently, huge progress has been made in the fields of DNA sequencing, X-ray crystallography and bioinformatics, which facilitate designing knowledge-based mutant libraries. For instance, information on the spots that may affect activity or thermostability can be obtained by aligning sequences among target enzymes and homologous enzymes. Multiple sequence alignment compares

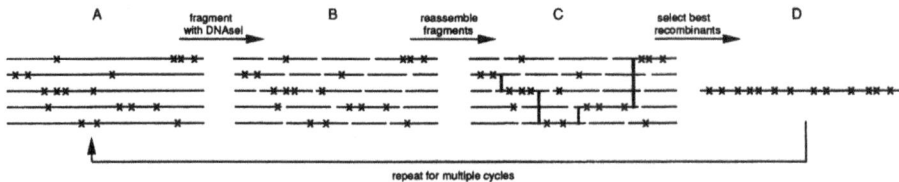

Scheme 2.11   Generation of chimeric gene libraries by DNA shuffling.[53]

similarities and differences in enzyme sequences based on a scoring matrix using a certain algorithm. Comparing the conserved and differing amino acids within the alignment result, also called a "consensus approach", sheds light on finding key spots in target enzymes for mutagenesis.

More importantly, structure-based design paves the way for deeper under-standing of structure–function relationships. The RCSB PDB databank (http://www.rcsb.org) provides a large number of experimentally determined protein structures or protein–ligand complexes, using X-ray diffraction, nuclear magnetic resonance or scanning electron microscope techniques. An enzyme–ligand complex structure envisions the binding mode of a substrate with an enzyme, in which the active site, intermolecular interactions and cat-alytic mechanism can be inferred. Sometimes enzyme structure is difficult to obtain. In this case, homology modeling can simulate the tertiary structure of a target enzyme using a homologous structure as a searching template. Numerous web servers and programs have been developed to build theoret-ical structures, including SWISS-MODEL, Robetta, 3D-JIGSAW, EsyPred3D, HHpred and Modeller. Molecular docking programs facilitate the viewing of enzyme–substrate interactions when the complex structure is not available experimentally. Compared to the determination of complex structures, these computational methods employ a rigid structure of active sites for energy cal-culations, which are time-efficient and can provide a reliable binding mode for protein design. To date, rational protein design is still a job of trial and error. In principle, researchers usually seek better substrate accommodation in the active binding cavity of an enzyme (by widening or prolonging the substrate entrance channel, or by reorganizing the cavity shape), stronger enzyme–sub-strate interactions (by adding π–π interactions, hydrogen bonds, disulfide bonds or salt bridges), faster product-release velocity (*e.g.*, by enlarging the product-release channel) or more rigid protein structure (by increasing hydro-gen bonds, disulfide bonds or salt bridges, or by modifying hydrophilic amino acids on the protein surface). Exchange with unnatural amino acids was also proved to be a new avenue of protein design for better catalytic properties.[57]

A structure-based rational or semi-rational protein design significantly reduces experimental library-building and mutant screening efforts whilst maximizing the percentage of improved clones with a given property. A typical approach, dubbed iterative saturation mutagenesis (ISM), employs successive rounds of saturation mutagenesis on chosen amino acids. A small group of amino acids was chosen within the active site of protein structure (*via* either experiments or homology modeling) in a combinatorial active site test (CAST-ing) for improving the catalytic activity or enantioselectivity,[58] since these cat-alytic properties are majorly determined by the amino acids buried inside the catalytic cavity. In the case of enhancing thermostability, a technique termed the B-factor iterative test (B-FIT) can be performed by choosing amino acids with higher B-factors (based on experimental X-ray data), which indicates higher thermal motion and positional disorder.[59] Saturation mutagenesis on these "unstable" amino acid residues is expected to be effective for higher ther-mostability. This approach introduces randomness into the spots with gener-alized and rationally selected standards. Furthermore, it controls the screening requirement by modulating the codon design, such as changing the NNK codon for saturation mutagenesis to an NDT codon, which codes 12 amino acids whilst covering all the types of residues including hydrophilic, hydrophobic, alkaline or acidic. Such a codon alteration significantly reduces the screening requirement of two-site mutagenesis from 3000 to less than 500.[60]

A comprehensive description of mutagenesis techniques and successful examples in the past decades would be a daunting work. Numerous new and enlightening techniques are piling up for accurate, fast and simple discovery of more efficient, versatile or robust enzymes. We would like to share some recent examples from our laboratory in the following sections to exhibit the prowess of amino acid substitutions for modulating enzyme activity, selectivity, and thermostability.

2.5.2 Improving Enzyme Activities for Bulky Substrates

Beta-adrenergic receptor blocking agents (β-blockers) are important building blocks for many widely used cardiovascular drugs. Among them, single-enantiomer β-blockers, usually (*S*)-enantiomers, are of great interest due to their stronger binding affinity for the β-adrenergic receptor. A robust epoxide hydrolase, *Bm*EH from *Bacillus megaterium* ECU1001, exhibited a very high activity in the kinetic resolution of *rac*-glycidyl phenyl ether (PGE), which was a model substrate for the initial screening. However, *Bm*EH showed much lower activities towards bulkier substrates, including precursors of moprolol (**2a**), SR-59230A (**3a**), aprenolol (**4a**), toliprolol (**7a**), xibenolol (**8a**) and propranolol (**9a**) (Scheme 2.12).

We first determined the crystal structures of *Bm*EH and *Bm*EH-POA, where POA is an amide analogue of substrate PGE. The complex structure revealed an active tunnel of *Bm*EH, including a substrate-access site, an epoxide-hydrolysis site, and a product-release site. Met145 and Phe128, located in the epoxide-hydrolysis site and the product-release site, are proposed to impair the *Bm*EH activity towards bulky substrates. Preliminary mutagenesis confirmed that both variants M145A and F128A exhibited an expanded active tunnel and therefore higher activities on NGE (**9a**). Semi-saturated mutagenesis was carried out on hot spots Met145 and Phe12. Considering the steric

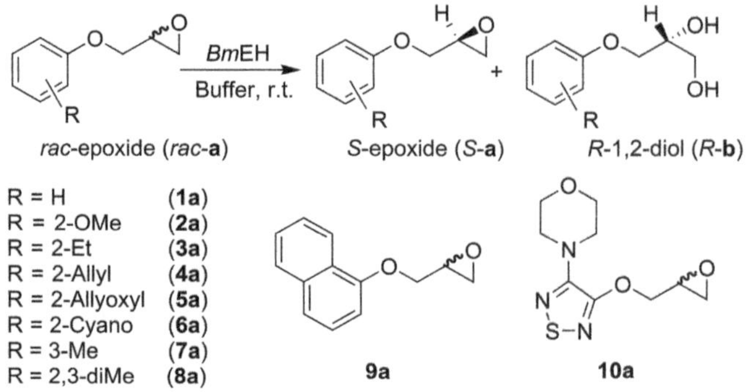
BmEH
Buffer, r.t.
R
rac-epoxide (rac-a)
S-epoxide (S-a)
R-1,2-diol (R-b)
R = H (1a)
R = 2-OMe (2a)
R = 2-Et (3a)
R = 2-Allyl (4a)
R = 2-Allyoxyl (5a)
R = 2-Cyano (6a)
R = 3-Me (7a)
R = 2,3-diMe (8a)
9a
10a

Scheme 2.12 Hydrolytic epoxide reactions driven by BmEH.

hindrance and hydrophobicity of amino acids, Met145 and Phe128 were mutated to seven amino acids (Ala, Cys, Ile, Leu, Ser, Thr and Val). Rationally preselecting mutation sites and limiting amino acid diversity creates a very smart mutant library with only 14 variants to screen. The ratio of improved variants was still extraordinary. Most variants showed an expanded substrate scope for bulky epoxide substrates (Figure 2.14). Generally, Met145 variants favored *ortho*-substituted substrates **3a** and **4a**, and Phe128 variants exhibited higher activities towards *meta*-substituted **7a** and **9a**. For instance, the kinetic constant of M145S against **3a** was 273-fold higher than that of the wild-type, while in the case of F128S against **9a** the improvement was as high as 896-fold. A preparative bioresolution of various epoxides using the corresponding best variants in the library resulted in 96.6-99.5% ee and 37-45% yields, which is appealing for manufacturing-scale production for β-blocker precursors.[61]

Enantiopure α-hydroxy acids are intermediates widely used for the synthesis of chiral drugs. For example, (*R*)-2-hydroxy-2-(2'-chlorophenyl)acetic acid is an important building block of (*S*)-clopidogrel, a popular drug for heart attack or stroke patients. A thermostable esterase rPPE01, isolated from *Pseudomonas putida* ECU1011, showed an excellent enantioselectivity (*E* > 200) toward α-acetoxycarboxylates, although the catalytic activity was still low. The structure of rPPE01 was initially modeled based on the crystal structures of the 3 most similar proteins (above 40% identity) using Modeller. Then the substrate molecule was docked into the active site of the modeled enzyme structure by AutoDock Vina program. Based on these bioinformatics and computational information, Trp187 and Asp287 were identified as key residues in determining activities, which were observed to coordinate with the alcohol

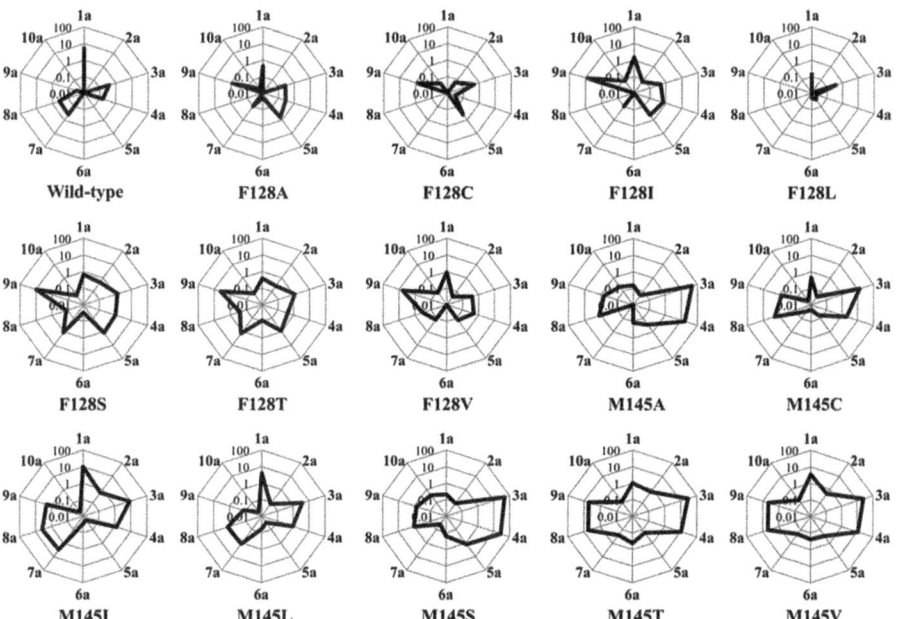

Figure 2.14   Specific activities of BmEH variants for various epoxides.

group and carboxyl group of the substrate, respectively (Figure 2.15). Two mutants (W187H and D287A) were designed to increase the polarity of carboxyl group binding pocket and enlarge the active site. As a result, both mutants displayed a significant improvement (100-fold) in catalytic efficiency for the substrate sodium 2-acetoxy-2-(2′-chlorophenyl)acetate (AcO-CPA-Na). In addition, W187H retained the excellent enantioselectivity ($E > 200$). More importantly, the increased polarity and size in the W187H active site made more efficient reactions with a series of α-acetoxycarboxylates possible, indicating that the esterase may become a potentially versatile biocatalyst for industrial applications[62] (Scheme 2.13). Further determination of crystal structures confirmed the mechanism of activity improvement,[63] suggesting that cyber methods such as homologous modeling and molecular docking might be a good prediction tool for mutagenesis design when crystal structure is not available.

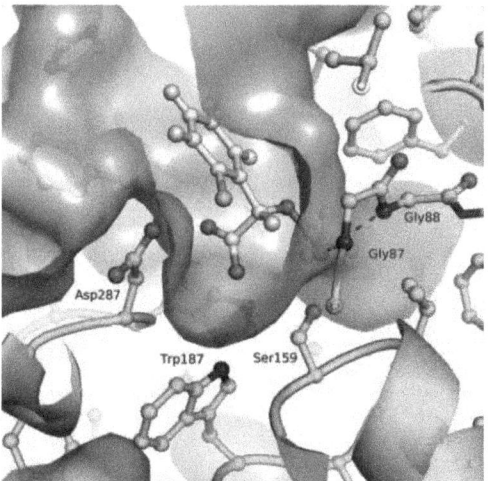

Figure 2.15   Docking pose of substrate (S)-Ac-CPA in the rPPE01 active site.

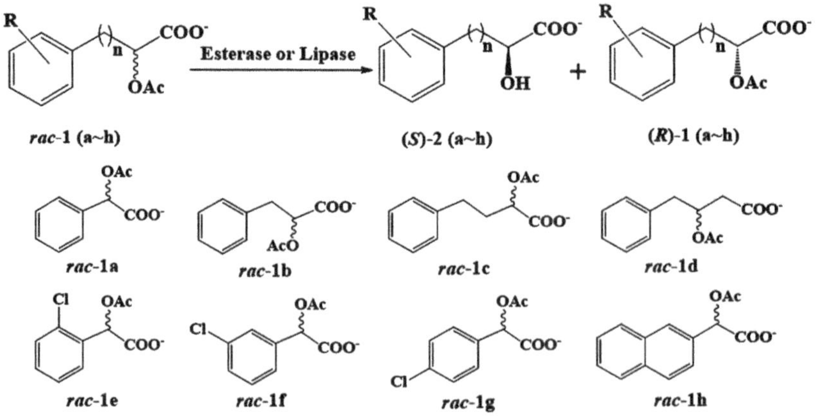

Scheme 2.13   Resolution of racemic α-acetoxycarboxylates using rRPPE01 variant.

2.5.3 Modulating Enzyme Selectivity for Preferable Substrates

A new *Po*OPH enzyme was identified from *Pseudomonas oleovorans* with high lactonase activity and latent organophosphorus hydrolase (OPH) activity. Numerous lactonases exist in various organisms and are involved in different physiological reactions. However, organophosphates (OPs), the most toxic synthetic compounds without natural analogs, are used as pesticides and chemical warfare agents. In order to enhance the capability of *Po*OPH to degrade toxic OPs, multiple sequence alignment was performed to compare *Po*OPH sequences with various homologous lactonases and OPHs. According to the consensus approach, two conserved residues His250 and Ile263 were chosen as key residues for determining substrate selectivity. Saturation mutagenesis was performed at the two positions, in which 20 amino acids were incorporated individually by the degenerated codon NNK to investigate the preferable residues for higher PTE activity. After combining the beneficial mutations, the resulting mutant *Po*OPH$_{M2}$ (H250I/I263W) exhibited an overall 6962-fold and 106-fold activity improvement against OP pesticides methyl-parathion and ethyl-paraoxon, respectively. On the contrary, the mutant displayed a substantial loss of activity for lactonase, switching the substrate preference from lactonase to phosphotriesterase (Scheme 2.14). Molecular docking revealed that the double mutations at the active site of *Po*OPH facilitated OP substrates accommodation.[64]

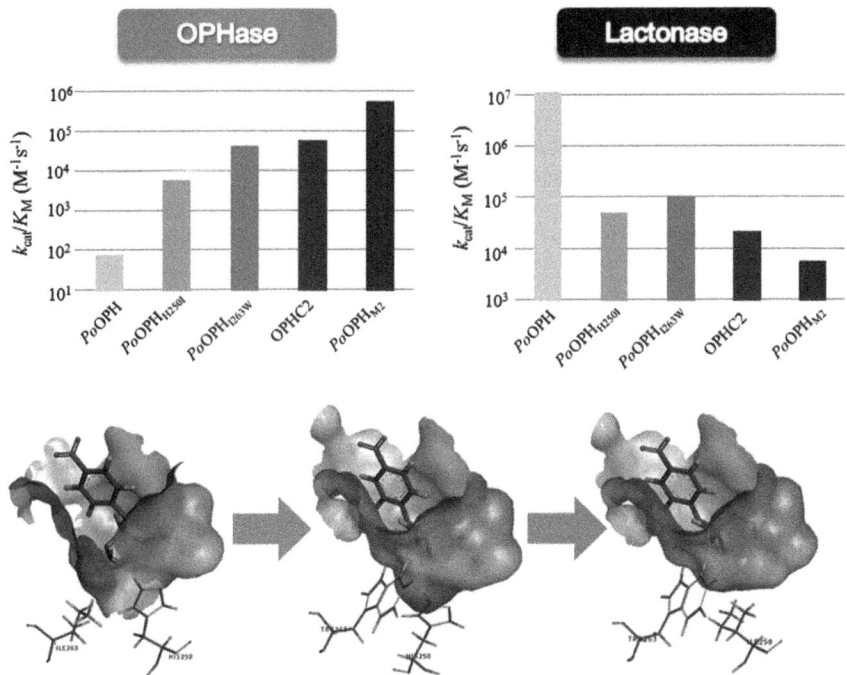

Scheme 2.14 Switching a lactonase into a phosphotriesterase by two amino acid substitutions.

Amino acid substitutions can also modulate the enantioselectivity or regioselectivity of enzymes. A typical example is the semi-rational mutagenesis of *Bacillus amyloliquefaciens* esterase (BAE) for higher enantioselectivity. (*R*)-1-(3′,4′-Methylenedioxyphenyl)ethanol is a valuable chiral intermediate for the synthesis of cancer drugs steganacin, podophyllotoxin or etoposide. Wild-type BAE showed only moderate enantioselectivity (*E* = 35). The preference for the (*R*)-substrate was still not satisfactory (*E* = 140) after decreasing the reaction temperature and adding surfactant.[28] Combinatorial active site testing (CASTing) was employed to further improve the enantioselectivity. Eight sites that comprise amino acids around catalytic Ser190 (8 Å) were chosen for site-directed saturation mutagenesis. The conserved catalytic residues were excluded from mutagenesis to avoid undermining the BAE activity. After performing mutagenesis iteratively, the beneficial mutations were accumulated to achieve a superior mutant with higher selectivity enhancement (ee > 97%, *E* = 186).[65]

2.5.4   Enhancing Thermostability for More Robust Enzymes

Thermostability is an important criterion for enzyme capabilities, since it determines the lifetime of usually expensive biocatalysts and the suitability for continuous industrial reaction process. Traditionally, thermostability enhancement by protein engineering could be achieved by random mutagenesis, since it was usually difficult to spot the determining residues for stability merely from the structure of a protein. Although usually done by trial and error, thermostability can also be improved rationally using the B-FIT method, a consensus approach or mutagenesis at hydrophilic amino acids on the protein surface. In the case of reductase *Cg*KR1, consensus approach was employed, in which the *Cg*KR1 sequence was compared with a series of thermostable reductases (sequence identity ranging from 31% to 62%) to find key residues for mutagenesis.[66] The T_{50} (ref. 15) value of the resulting variant M3 was 2.3 °C higher compared to that of the wild-type enzyme. The variant offered a very good starting point for significantly enhancing the protein stability by directed evolution.

References

1. http://en.wikipedia.org/wiki/File:Chirality_with_hands.svg.
2. A. M. Thayer, *Chem. Eng. News*, 2007, **85**, 11.
3. J. S. Carey, D. Laffan, C. Thomson, *et al.*, *Org. Biomol. Chem.*, 2006, **4**, 2337.
4. R. Noyori, 2001, http://nobelprize.org/nobel_prizes/chemistry/laureates/2001/noyori-lecture.pdf.
5. K. Tani, T. Yamagata, S. Akutagawa, *et al.*, *J. Am. Chem. Soc.*, 1984, **106**, 5208.
6. R. Goswami, *J. Am. Chem. Soc.*, 1980, **102**, 5974.

7. J. M. Woodley, *Trends Biotechnol.*, 2008, **26**, 321.
8. H. Matsumae, M. Furui and T. Shibatani, *J. Ferment. Bioeng.*, 1993, **75**, 93.
9. D. H. Peterson, H. C. Murray, S. H. Eppstein, *et al.*, *J. Am. Chem. Soc.*, 1953, **74**, 5933.
10. B. Martin-Matute and J. E. Backvall, *Curr. Opin. Chem. Biol.*, 2007, **11**, 226.
11. H. Pellissier, *Tetrahedron*, 2008, **64**, 1563.
12. B. Schnell, K. Faber and W. Kroutil, *Adv. Synth. Catal.*, 2003, **345**, 653.
13. S. Luetz, L. Giver and J. Lalonde, *Biotechnol. Bioeng.*, 2008, **101**, 647.
14. O. May, *Biocatalysis for the Pharmaceutical Industry*, ed. J. H. Tao, G. Q. Lin and A. Liese, John Wiley & Sons (Asia) Pte Ltd, Singapore, 2008, ch. 14, pp. 305–321.
15. D. J. Pollard and J. M. Woodley, *Trends Biotechnol.*, 2007, **25**, 66.
16. J. Pan, H. L. Yu, J. H. Xu and G. Q. Lin, *Organomet. Chem.*, 2011, **36**, 67.
17. G. W. Zheng and J. H. Xu, *Curr. Opin. Biotechnol.*, 2011, **22**, 784.
18. Y. Ni and J. H. Xu, *Biotechnol. Adv.*, 2012, **30**, 1279.
19. A. J. J. Straathof, S. Panke and A. Schmid, *Curr. Opin. Biotechnol.*, 2002, **13**, 548.
20. D. Shen, J. H. Xu, P. F. Gong, *et al.*, *Can. J. Microbiol.*, 2001, **47**, 1101.
21. D. Shen, J. H. Xu, H. Y. Wu, *et al.*, *J. Mol. Catal. B: Enzym.*, 2002, **18**, 219.
22. P. F. Gong, H. Y. Wu, J. H. Xu, *et al.*, *Appl. Microbiol. Biotechnol.*, 2002, **58**, 728.
23. Y. Y. Liu, J. H. Xu, Q. G. Xu and Y. Hu, *Biotechnol. Lett.*, 1999, **21**, 143.
24. Y. Y. Liu, J. H. Xu and Y. Hu, *J. Mol. Catal. B: Enzym.*, 2000, **10**, 523.
25. T. W. Xu and J. H. Xu, *Biotechnol. J.*, 2006, **1**, 1293.
26. H. Y. Wu, J. H. Xu and Y. Y. Liu, *Synth. Commun.*, 2001, **31**, 3491.
27. Y. Y. Liu, J. H. Xu, H. Y. Wu, *et al.*, *J. Biotechnol.*, 2004, **110**, 209.
28. J. Y. Liu, G. W. Zheng, *et al.*, *Biotechnol. Bioprocess Eng.*, 2014, **19**, 442.
29. Y. F. Tang, J. H. Xu, Q. Ye and B. Schulze, *J. Mol. Catal. B: Enzym.*, 2001, **13**, 61.
30. P. F. Gong, J. H. Xu, Y. F. Tang and H. Y. Wu, *Biotechnol. Prog.*, 2003, **19**, 652.
31. J. Zhao, Y. Y. Chu, A. T. Li, X. Ju, J. Pan, Y. Tang and J. H. Xu, *Adv. Synth. Catal.*, 2011, **353**, 1510.
32. P. F. Gong and J. H. Xu, *Enzyme Microb. Technol.*, 2005, **36**, 252.
33. W. Xu, J. H. Xu, J. Pan, *et al.*, *Org. Lett.*, 2006, **8**, 1737.
34. Q. Q. Zhu, W. H. He, X. D. Kong, *et al.*, *Appl. Microbiol. Biotechnol.*, 2013, **98**, 207.
35. W. J. Chen, W. Y. Lou, C. Y. Yu, *et al.*, *J. Biotechnol.*, 2012, **162**, 183–190.
36. J. H. Su, J. H. Xu, H. L. Yu, *et al.*, *J. Mol. Catal. B: Enzym.*, 2009, **57**, 278.
37. J. H. Su, J. H. Xu and Z. L. Wang, *Appl. Biochem. Biotechnol.*, 2010, **160**, 1116.
38. A. M. Tong, W. Y. Lu, J. H. Xu and G. Q. Lin, *Bioorg. Med. Chem. Lett.*, 2004, **14**, 2095.
39. A. M. Tong, J. H. Xu, W. Y. Lu and G. Q. Lin, *J. Mol. Catal. B: Enzym.*, 2005, **32**, 83.
40. H. L. Yu, J. H. Xu, W. Y. Lu and G. Q. Lin, *J. Biotechnol.*, 2008, **133**, 469.

41. H. L. Yu, J. H. Xu, Y. X. Wang, W. Y. Lu and G. Q. Lin, *J. Comb. Chem.*, 2008, **10**, 79.
42. R. L. Yang, N. Li and M. H. Zong, *J. Mol. Catal. B: Enzym.*, 2012, **74**, 24.
43. W. W. Xi and J. H. Xu, *Process Biochem.*, 2005, **40**, 2161.
44. L. L. Zhao, J. Pan and J. H. Xu, *Biotechnol. Bioprocess Eng.*, 2010, **15**, 199.
45. X. Zhang, J. H. Xu, D. H. Liu, J. Pan and B. Chen, *Biochem. Eng. J.*, 2010, **50**, 47.
46. B. Hu, J. Pan, H. L. Yu, J. W. Liu and J. H. Xu, *Process Biochem.*, 2009, **44**, 1019.
47. H. Zhu, J. Pan, B. Hu, H. L. Yu and J. H. Xu, *J. Mol. Catal. B: Enzym.*, 2009, **61**, 174.
48. Q. Dong, L. M. Ouyang, H. L. Yu and J. H. Xu, *Carbohydr. Res.*, 2010, **345**, 1622.
49. G. W. Zheng, H. L. Yu, C. X. Li, J. Pan and J. H. Xu, *J. Mol. Catal. B: Enzym.*, 2011, **70**, 138.
50. J. Pan, X. D. Kong, C. X. Li, Q. Ye, J. H. Xu and T. Imanaka, *J. Mol. Catal. B: Enzym.*, 2011, **68**, 256.
51. C. A. Tracewell and F. H. Arnold, *Curr. Opin. Chem. Biol.*, 2009, **13**, 3.
52. N. J. Turner, *Nat. Chem. Biol.*, 2009, **5**(8), 567.
53. W. Stemmer, *Proc. Natl. Acad. Sci. U. S. A.*, 1994, **91**, 10747.
54. N. Tokuriki, C. J. Jackson, L. Afriat-Jurnou, K. T. Wyganowski, R. Tang and D. S. Tawfik, *Nat. Commun.*, 2012, **3**, 1257.
55. N. Tokuriki and D. S. Tawfik, *Nature*, 2009, **459**, 7247.
56. R. Ostafe, R. Prodanovic, J. Nazor and R. Fischer, *Chem. Biol.*, 2014, **21**, 414.
57. C. C. Liu and P. G. Schultz, *Annu. Rev. Biochem.*, 2010, **79**, 413.
58. M. T. Reetz, L. W. Wang and M. Bocola, *Angew. Chem., Int. Ed.*, 2006, **118**, 1258.
59. M. T. Reetz, J. D. Carballeira and A. Vogel, *Angew. Chem., Int. Ed.*, 2006, **45**, 7745.
60. M. T. Reetz and J. D. Carballeira, *Nat. Protoc.*, 2007, **2**(4), 891.
61. X. D. Kong, Q. Ma, J. H. Zhou, B. B. Zeng and J. H. Xu, *Angew. Chem., Int. Ed.*, 2014, **53**(26), 6641.
62. B. D. Ma, X. D. Kong, H. L. Yu, Z. J. Zhang, S. Dou, Y. P. Xu, Y. Ni and J. H. Xu, *ACS Catal.*, 2014, **4**(3), 1026.
63. S. Dou, X. D. Kong, B. D. Ma, Q. Chen, J. Zhang, J. H. Zhou and J. H. Xu, *Biochem. Biophys. Res. Commun.*, 2014, **446**, 1145.
64. X. J. Luo, X. D. Kong, J. Zhao, Q. Chen, J. H. Zhou and J. H. Xu, *Biotechnol. Bioeng.*, 2014, **111**(10), 1920.
65. J. Y. Liu, H. P. Bian, Y. Tang, Y. P. Bai and J. H. Xu, *Appl. Microbiol. Biotechnol.*, 2015, **99**, 1701.
66. L. Huang, H. M. Ma, H. L. Yu and J. H. Xu, *Adv. Synth. Catal.*, 2014, **356**, 1943.

CHAPTER 3

Regulation of Non-canonical DNA Structures by Small Molecules and Carbon Materials

CHONG WANG[a], JINGYAN ZHANG*[a], AND SHOUWU GUO*[b]

[a]State Key Laboratory of Bioreactor Engineering, Shanghai Key Laboratory of New Drug Design, School of Pharmacy, East China University of Science and Technology, Shanghai, 200237, P. R. China; [b]Research Institute of Micro/Nano Science and Technology, Shanghai Jiao Tong University, Shanghai, 200240, P. R. China
*E-mail: jyzhang@ecust.edu.cn, swguo@sjtu.edu.cn

3.1 Introduction

The regular double-helix structure of DNA, typically called B-DNA, in which two complementary strands are held together by Watson–Crick base pairs, is well recognized. Recently it has been found that under certain conditions DNA can form non-canonical structures, such as Z-DNA, A-motif, G-quadruplex, i-motif, triplex, hairpin, and cruciform.[1-8] These structures are particularly seen in the human genome with repeat DNA sequences, which account for more than 50% of the total genomic DNA, while simple sequence repeats account for approximately 3% of the total DNA. Although some of these unusual DNA structures have not been directly detected *in vivo*, they have been

RSC Green Chemistry No. 34
Chemical Biotechnology and Bioengineering
By Xuhong Qian, Zhenjiang Zhao, Yufang Xu, Jianhe Xu, Y.-H. Percival Zhang, Jingyan Zhang, Yangchun Yong, and Fengxian Hu

proposed to participate in several biologically important processes.[9–14] Some of the non-canonical DNA is expected to act as a signpost for oncogenes and a controller for oncogene expression at the transcription level in the future because non-canonical DNA sequences are frequently observed in and near the promoter regions of oncogene and human telomeric DNA.[15] Since they are prone to structural alterations, non-canonical DNA structures are hot spots for chromosomal breaks, homologous recombination, and gross chromosomal rearrangements, thus could be potential drug targets.[9–14] It has actually been found that non-canonical DNA is associated with human diseases such as myotonic dystrophy, Huntington's disease, fragile X syndrome, and Friedreich's ataxia.[16–20] Therefore, finding out how to regulate these structures is of importance in terms of gene regulation, molecular recognition, and drug discovery. Due to the unique three dimensional structures of non-canonical DNA molecules, small molecules that bind to them can stabilize or alter their structures, and thus are potentially able to regulate their biological functions. In the current chapter we have included a brief introduction of several typical non-canonical DNA structures, then focused on the regulation of these non-canonical DNA structures with small molecules and carbon materials.

3.1.1 Non-canonical DNA Structures

More than 10 different types of non-canonical DNA structure have been reported so far, including Z-DNA, A-motif, tetraplex (G-quadruplex and i-motif), hairpin, cruciform, and triplex, and their structures have been described in several reviews.[2,5,6] Here, five typical non-canonical DNA structures will be briefly introduced.

3.1.1.1 Z-DNA

As its name indicates, the phosphate backbone of Z-DNA has a zigzag pattern and, in contrast to typical B-DNA, it is a left-handed double helix; its diameter is approximately 18 Å (~20 Å for B-DNA), and there are 11.6 base pairs per helical turn compared to 10.5 base pairs for B-DNA. The glycosidic bonds of the bases have alternating *syn*- and *anti*-conformations.[2,21] Thus the bases of Z-DNA are relatively far away from the axis, and a single narrow groove exists instead of the major and minor grooves observed in B-DNA. In fact, it is so narrow that nothing penetrates the groove other than water molecules, not even hydrated ions. The groove is also appreciably narrower than either the major or the minor groove of right-handed B-DNA, but it is about 9 Å deep. As only purine residues can adopt the *syn* conformation, Z-DNA is found in alternating purine–pyrimidine sequences; $(GC)_n$ is the most favored sequence for Z-DNA. Interestingly, $(GC)_n$ sequences are typical sequences in telomeric DNA as well as being a promoter region in many oncogenes, suggesting that the formation of Z-DNA may be related to carcinogenesis, which arouses the interest of scientists.[22] The unique structural features of Z-DNA may account for the interactions of Z-DNA with macromolecule and small molecules.

3.1.1.2   A-motif DNA

A-motif DNA, also called poly(dA), is a single-stranded right-handed helical structure stabilized by the π–π stacking of adenine bases at alkaline and neutral pH. At acidic pH, poly(dA) forms a right handed helical duplex with parallel-mannered chains and tilted protonated bases. It is believed that that the parallel duplex is stabilized by two factors: the hydrogen bonds (reverse Hoogsteen base-pairing, see Figure 3.1) between two protonated adenine bases and electrostatic attraction between the positively charged protons at the N_1 atom of the adenines and the negatively charged phosphate groups.[23,24] Various spectroscopic studies have shown that different forms of poly(dA) exist under different conditions. The structure of poly(A) or poly(dA) can be tuned by the pH of the environment or by small molecules.

3.1.1.3   DNA Triplex

A DNA triple helix contains three strands, two of the DNA double helix, with the third strand winding around the major groove of the DNA double helix forming a triplex. There are two types of triplexes, inter-molecular and intra-molecular, depending on the source of the third

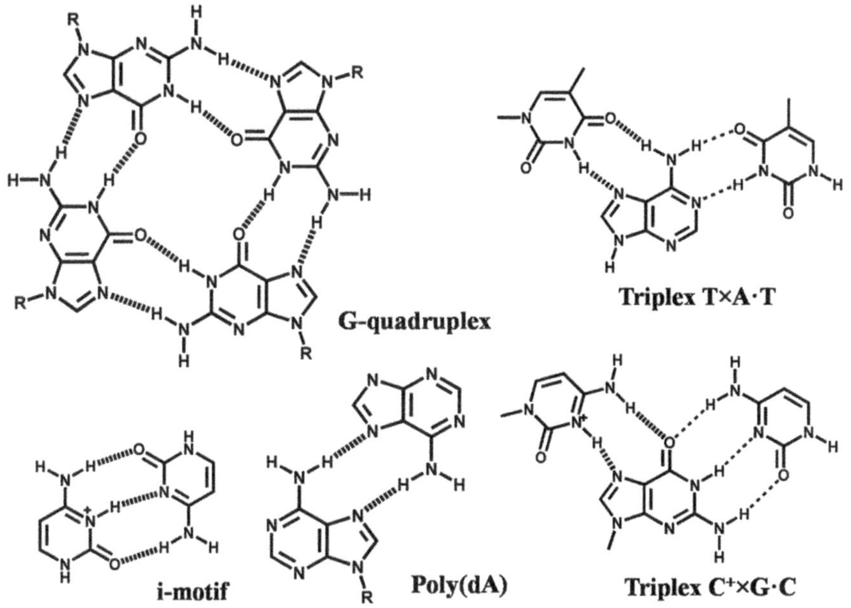

Figure 3.1   Hydrogen bond formation in G-tetrad, parallel triplexes consisting of T × A·T and C⁺ × G·C triads, A-motif, and i-motif (Watson–Crick base-pairing is shown with dashed bonds, and Hoogsteen or reversed Hoogsteen base-pairing is shown with hashed bonds).

strand. Inter-molecular triplexes are formed when the triplex-forming strand is a separate single strand containing deoxyribonucleotide or one of the strands from a different DNA molecule, while intra-molecular triplexes are formed when DNA with polypyrimidine/polypurine sequences with mirror symmetry undergoes conformational rearrangement and folds back onto the duplex itself. The third strand is usually rich in polypurine/polypyrimidine. The binding of the third strand in the triplex is through Hoogsteen or reverse-Hoogsteen type hydrogen bonding, which is different from the classical Watson–Crick base pairing of B-DNA as shown in Figure 3.1. Depending on the bases (pyrimidine or purine) of the third strand interacting with the purine base of the duplex, there are two kinds of DNA triplex: parallel and anti-parallel, respectively. The most well-known triads of the parallel triplex are T–A:T and C^+–G:C, whereas the anti-parallel triplex consists of G–G:C and A–A:T triads.[6] One easy method to determine the formation of a DNA triplex is through UV melting experiments; low and high melting temperatures can be observed, corresponding to the dissociation of one strand from the DNA duplex and dissociation of the DNA duplex, respectively. The rise of the triplex is about 3.2 ± 0.2 Å, and the twist is in the range of $34 \pm 1.6°$, 10.5 base pairs per turn, which is larger at the duplex end than in the triplex region. In the region of the duplex where the third strand interacts, the minor groove is highly compressed, reaching a minimum of 3.5 Å (6–7 Å in standard B DNA), whereas in the duplex extensions the groove width is wider and more similar to that of standard B DNA. However, the minor groove at the junction is considerably wider (8.5 Å) than elsewhere.[25] As a result of the third strand binding in the DNA duplex, the major groove of the triplex is divided into two parts: the Crick–Hoogsteen groove and the Watson–Hoogsteen groove. The groove widths are essentially dependent of sequence and essential to its stability and recognition.

3.1.1.4   G-quadruplex

G-quadruplex DNA is formed by G-rich DNA sequences; it has at least two contiguous G-tetrads stacked one on top of the other and is stabilized by monovalent cations (K^+, Na^+, NH^+, and Pb^{2+}) that are coordinated to the lone pairs of electrons of O_6 in each guanine.[26-30] The hydrogen bonds between four guanines are through Hoogsteen-type base pairing mode. G-quadruplexes can be formed by intramolecular folding of G-rich sequences or by intermolecular association of two or four sequences (formation of dimeric or tetrameric quadruplexes). Due to the nucleotide sequences, number and orientation of the strands, the *syn/anti* glycosidic conformation of guanines, the length of the strands and loops, and environmental factors such as cations that ae used to stabilize them, G-quadruplexes display a wide range of structural polymorphisms, which have been described in review articles.[31,32] Among all the structural features, the most prominent is their larger planar G-tetrad.

3.1.1.5   i-Motif DNA

i-Motif DNA is formed by two parallel-stranded $C:C^+$ hemiprotonated base-paired duplexes that are intercalated in an antiparallel manner. i-Motifs are classified into two classes based on the length of their loops. Class I i-motifs have shorter loops, whereas class II i-motifs have longer loops. The i-motif structure is formed at slightly acidic or even neutral pH[33] and its structure is significantly affected by the number of cytosine bases it contains. i-Motif DNA is structurally dynamic over a wide pH range, adopting multiple conformations ranging from the folded i-motif structure to a random coil conformation based on theoretical simulation and small-angle X-ray techniques.[34] Similar to G-quadruplex DNA, i-motif DNA also shows a high degree of structural polymorphism depending on the number of cytosine bases,[35] loop length,[36] environmental condition,[37,38] and attached or interacting material with the DNA strands.[39–41] In general, class II i-motifs are reasoned to be the more stable due to extra stabilizing interactions within the longer loop regions. The typical i-motif core includes various $C \cdot C^+$ pairs, which are nearly planar with deviations less than 18°. The average stacking interval is 3.1 Å. The average π–π distance is 14.2 Å across the wide grooves and 9.4 Å across the narrow grooves.[42]

3.2   Regulating the Structures of Non-canonical DNA with Small Molecules

Non-canonical DNA structures are frequently observed in and near the promoter regions of oncogene and human telomeric DNA, thus they have potential in disease detection and as drug targets. The biological functions of non-canonical DNA are significantly implicated in their structures and stabilities. As summarized in Section 3.1, each category of non-canonical DNA structures has its own unique structural features, thus molecules that can bind and react with DNA at specific sites provide a means of regulating their stability, structure, and functions. Generally, there are two types of interaction modes between small molecules and nucleic acids: covalent and non-covalent binding. The latter mode is more often seen and generally includes intercalation, groove binding, stacking along the DNA helical axis, and DNA terminus capping. The interaction of small molecules with B-DNA has been well documented.[43] Here we primarily focus on the non-covalent interactions of small molecules with non-canonical DNA structures.

3.2.1   Small Molecules that Induce and Stabilize Z-DNA Structures

Z-DNA, containing alternating purine and pyrimidine repeats, is one of the most well studied structures besides A- and B-DNA. As described in Section 3.1.1.1, Z-DNA is a left-handed helix in which the purines are in the *syn* conformation and the pyrimidines in the *anti* conformation. Hence, it can be recognized due to its geometrical features and, to a lesser extent, by the sequences of the nucleobases. Unlike the i-motif structure, Z-DNA exists *in vivo* under physiological conditions as a transient structure occasionally induced by a biological process, such as transcription, the methylation of cytosine, and the level of DNA supercoiling. Z-DNA can be stabilized *in vivo* by negative super coiling, mutations, Z-DNA binding proteins, spermine, and spermidine. Therefore, organic ligands and chiral metal complexes have been used as Z-DNA promoters *in vitro*. For instance, Tsuji *et al.* demonstrated the efficient synthesis of a chiral Z-DNA binding ligand by the conjugation of a helicene unit with a spermine unit.[44] The cationic spermine portion produced electrostatic interactions along the phosphate backbone of the minor groove of Z-DNA and the helicene formed complexes in an end-stacking mode with Z-DNA. This binding mode, together with the thermodynamic parameters, accounted for the mode of chiral recognition of (P)- and (M)-3 for B- and Z-DNA. The results provided useful knowledge for designing small molecular ligands for the recognition of B- or Z-DNA.[44]

The majority of Z-DNA interacting small molecules are metallocomplexes, which might be due to their versatile structural properties.[5] Spingler *et al.* reported that copper complexes are likely to induce the stabilization of Z-DNA, while zinc complexes are not, even with the same structure.[45] They also showed that even with small modifications on the ligand, the metal complexes behaved markedly differently. It was assumed that the bisfluoro substitution on the ligand was causing the complex to be in the neutral coordination mode at the experimental conditions of pH 7, thus the electrostatic contribution together with the shielding effect of the ligand might explain the absence of any interaction with the DNA.[45] $[Ru(dip)_2dppz]^{2+}$ (dip = 4,7-diphenyl-1,10-phenanthroline, dppz = dipyridophenazine) can efficiently induce the B to Z transition of various DNA sequences, such as non-APP and full-AT sequences.[46] Though the thermodynamics and kinetics of the transition from B- to Z-DNA of various DNA were monitored using ITC and real-time CD spectroscopy, and could even be monitored in living HeLa cells, the functional mechanism of the $[Ru(dip)_2dppz]^{2+}$ remains not fully understood. Choi *et al.* investigated the chiroptical properties and binding modes of anionic and cationic free-base porphyrins and their nickel(II) and zinc(II) derivatives with right- and left-handed alternating cytosine–guanine poly$(dG–dC)_2$ DNA sequences using CD and absorption spectroscopies.[47] As previously reported, interaction of cationic porphyrins (2HT4, ZnT4, NiT4) with B-DNA caused large red shifts and hypochromisms of the Soret band and induced negative CD Cotton effects, suggesting the transition of B-DNA to A-DNA. In contrast, no interactions exist between anionic porphyrins and B-DNA. Spectroscopic data suggested that 2HT4 and NiT4 intercalated to Z-DNA, whereas ZnT4 bound to the outside of Z-DNA stereoselectively. Interestingly, they found that NiTPPS was able to spectroscopically discriminate between spermine-induced Z-DNA and Co(III)-induced Z-DNA *via* a new induced circular dichroism signal in the visible region of the electromagnetic spectrum (Figure 3.2).[48] They believed that not only the small molecules but also the Z-DNA inducers played an important role in the porphyrin–DNA binding mode.

Medina-Molner *et al.* reported that a dinuclear metallocomplex consisting of two 1,5,9-triazacyclododecane ligands linked by a propylene moiety, rather than the corresponding mononuclear metal complexes, could induce the formation of Z-DNA from B-DNA based upon the geometrical differences between the two DNA forms (Figure 3.3). The authors demonstrated the requirements for the dinuclear complexes to have such ability, including: metal-to-metal distance when bound to Z-DNA should be between 5 and 7 Å;

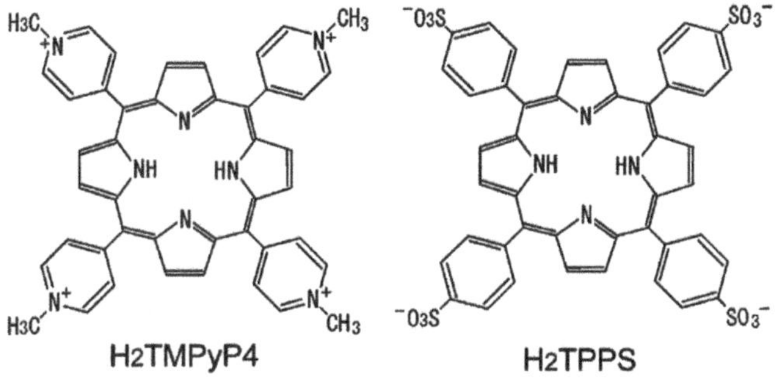

Figure 3.2 Structural formulae of porphyrins.[48] Reprinted with permission from ref. 48. Copyright 2005 American Chemical Society.

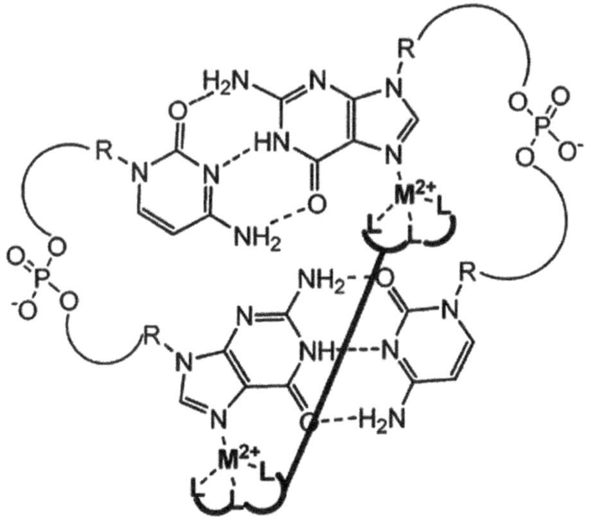

Figure 3.3 Proposed Z-DNA recognition by dinuclear metal complexes that bind to N(7) atoms of the two guanine bases of a d(GC)·d(GC) dimer.[49] Reprinted with permission from ref. 49. Copyright 2012 The Royal Society of Chemistry.

a flexible ligand backbone should allow optimal coordination to the DNA bases; and each metal center should have at least one vacant or labile coordination site. The cooperative effect of dinuclear centers might be used as a general principle of DNA recognition by metal complexes.[49]

In addition to the metallocomplexes, metal cations, primarily divalent cations, which tend to take coordination structures specific to Z-DNA, could thus stabilize its structure. Indeed, many crystal structures of Z-DNA have been determined with Mg^{2+}, Ba^{2+}, Cu^{2+}, and Mn^{2+}.[50–53] These structures showed that most of the identified Mg^{2+} cations interacted indirectly with Z-DNA *via* water molecules, or with the oxygen atoms of phosphate groups and the guanine N_7 atom of Z-DNA. Recently, it was reported that Z-DNA–metal complexes could be formed even at high concentrations of alkaline earth salts (500 mM).[54] Mg^{2+} and Ca^{2+} cations in these structures interact directly with the phosphate groups of Z-DNA duplexes instead of forming water-mediated interactions with Z-DNA in the presence of lower concentrations of alkaline earth salts. In the crystals of Z-DNA–metal complexes, Z-DNA was laid along its *c*-axis and interacted with its 6 neighboring DNA duplexes through coordination bonds of P–O⋯(Mg^{2+} or Ca^{2+})⋯O–P. A symmetrical hexagonal Z-DNA duplex assembly model may explain DNA condensation caused by alkaline earth salts. These structures offer insights into the functions of alkaline earth cations, which are essential to the structures and assembly of Z-DNA duplexes.

3.2.2 Small Molecules that Bind and Modulate A-motif Structures

As described in Section 3.1.1.2, A-motif DNA exhibits a single-stranded helical structure stabilized by the π–π stacking of adenine bases at alkaline and neutral pH, and a right-handed helical duplex with parallel-mannered chains and tilted protonated bases at acidic pH. The duplex poly(A) can be stabilized by alkaloids, such as berberine, palmatine, sanguinarine, and coralyne, which have been well reviewed.[55] Among them, coralyne has been well studied as an anti-leukemic agent whose anticancer activity is likely related to its ability to intercalate DNA; it also stabilizes DNA in the topoisomerase complex and inhibits the activity of this enzyme.[56,57] Coralyne was found to bind to A-motif DNA with a stoichiometry of one coralyne per four adenine bases, and to be released from the A-motif structure in a cooperative melting transition around 50 °C (Figure 3.3).[58] Further research suggested that coralyne promotes the formation of an antiparallel homo-adenine duplex at neutral pH. Furthermore, the helix of this duplex is compatible with flanking Watson–Crick helices. It was also found that the homoadenine–coralyne structure could be incorporated into duplex structures that include Watson–Crick pairs. Joung *et al.* modeled the characteristic structure of the coralyne-induced homo-(dA) duplex using molecular dynamics simulations with both explicit and implicit solvent.[59] In order to predict a structure, a hierarchical approach was applied building tentative coralyne-free duplexes, followed by single coralyne bound and multiple coralyne bound structures, paring down the list of proposed structures at each stage. Proposed structures included models with all possible A·A base pairing types. The most favorable structure in terms of energy, structural stability, and hydrogen bonding was the *trans* WH geometry, as shown in Figure 3.4. The simulation result was validated by the experimental data. Interestingly, Çetinkol and coworkers later reported that coralyne–poly(A) binding produced very sharp transitions between single-stranded and duplex structures as a function of coralyne concentration from dilution experiments, suggesting that the coralyne–poly(A)complex begins to dissociate under a certain critical concentration.[60] The results revealed that small molecules can be used to alter the secondary structure of A-motif DNA, and further small molecule binding can even be used to drive the formation of non-Watson–Crick A-motif structures.

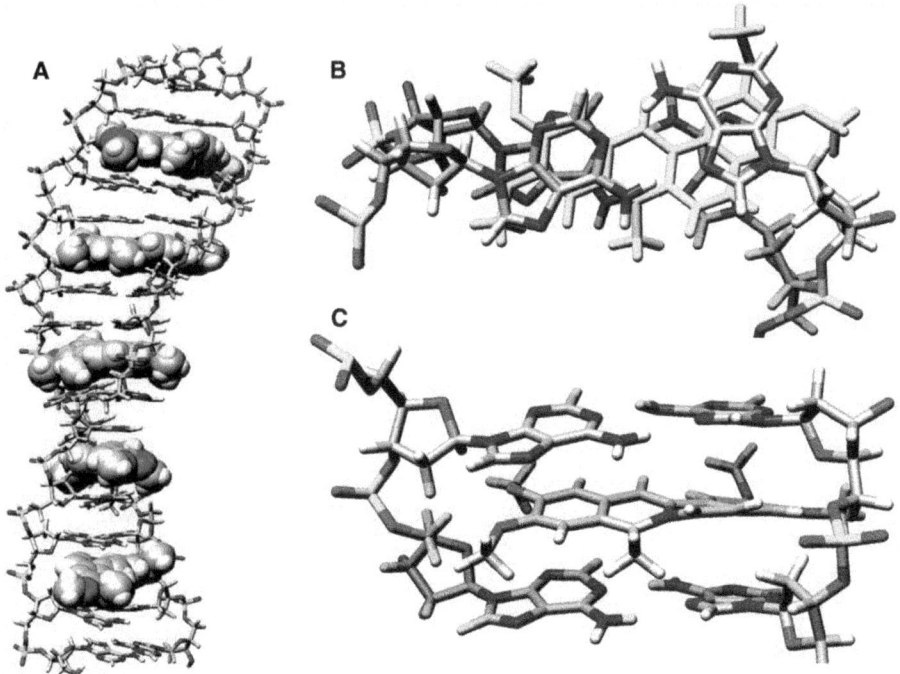

Figure 3.4   Graphical illustrations of representative structures of the *trans* WH geometry from cluster analysis. (A) The entire coralyne-induced homo-(dA) duplex with the DNA shown as a stick figure and the coralyne as solid spheres. (B) View of the stacking with coralyne (yellow) for the minimal binding unit. (C) A close-up view of the minimal binding unit. The kink in the coralyne structure at the nitrogen imparted by the nearby methyl leads to a distinct and slight buckle in the W–H base pairs.[59] Reprinted with permission from ref. 59. Copyright 2009 Oxford University Press.

3.2.3 Small Molecules that Stabilize Triplex DNA Structures

DNA triplexes consist of the Watson–Crick base pair and the Hoogsteen base pair together, and they are stabilized by hydrogen ions and preferentially formed under acidic conditions. Thus, small molecules bind to the triplex through two modes, intercalation or groove binding, based on their structural and physical features. Due to its unique structural properties, the required properties of a triplex-specific ligand have been summarized.[61] It was found that the small molecules should: be cationic, although the positive charge may be unfavorable for the stabilization of $C^+ \cdot G \cdot C$ triplets due to charge repulsion; have an aromatic surface; and also have an unfused aromatic system with torsional flexibility.

Most of the compounds that can interact with DNA triplexes are intercalators, such as acridine and its derivatives. Acridine was proposed to intercalate at the triplex–duplex junction, which is thought to be a strong intercalation site. The general structures of such compounds are shown in Figure 3.5.[62] Benzopyridoindoles and benzopyridoquinoxalines were the first molecules reported to bind to triplex DNA more tightly than that with duplex B-DNA; the positively charged side chains in these complexes might be located in the DNA grooves.[63] Several other intercalators have been reported to stabilize triplex DNA, such as coralyne,[64,65] naphthylquinoline compounds,[66–68] and disubstituted anthraquinones.[68–73] To improve the stabilizing efficiency of the small molecules, homopyrimidine ODN linked with an intercalator and acridine derivatives forming ODN–intercalator conjugates were also prepared.[74,75] The third oligonucleotide strand itself can also serve as a molecule to stabilize DNA triplexes by linking to a chemically reactive group of the triplex. Azidoproflavin or azidophenacyl have been frequently used to form monofunctional adducts.[62]

The other large category of small molecules that can stabilize DNA triplexes is groove binders, which have been summarized recently.[76] As triplex formation results from the binding of a third DNA strand within the major groove of the host duplex, this major groove is thus divided into two asymmetric grooves, a major groove and a minor groove, although the host minor groove remains essentially unaltered. These three grooves provide different binding sites to small molecules because each of them has a different nature and dimensions: even the two minor grooves are not identical. The interaction of the groove binder and triplex DNA is either through hydrogen bonding or electrostatic interaction, thus the binding of the groove binder to triplex DNA is pH and salt dependent. Groove binding ligands for triplex DNA are similar to the groove binders for B-DNA, such as DAPI, Hoechst dyes, distamycin, methyl green, Berenil, netropsin, terbenzimidazoles, mithramycin, and polyamines. Thus, they are not triplex-specific (Berenil, distamycin, Hoechst dyes), and some are even known to destabilize triplex DNA (berenil, distamycin) because of their preference for duplex DNA. Generally, the molecules that can simultaneously bind to the major and minor grooves, or the minor groove and the adjacent minor groove of the duplex, are highly effective triplex stabling reagents.[77,78]

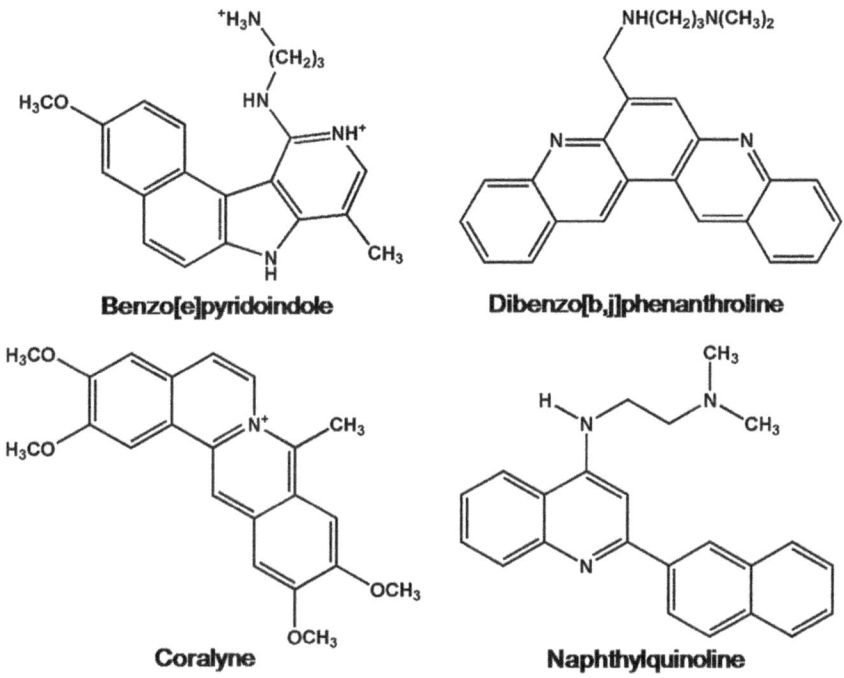

Figure 3.5 Typical DNA triplex-binding small molecules.

Arya *et al.* also summarized a new type of groove recognition of triplexes *via* amino sugars.[79] Among the reported amino sugars, neomycin is the most effective aminoglycoside for stabilizing a DNA triple helix. Neomycin selectively stabilizes triplex DNA (TAT and mixed base) without any effect on the DNA duplex. The selectivity of neomycin likely originates from its potential and shape complementarity to the triplex Watson–Hoogsteen groove, making it the first molecule that selectively recognizes a triplex groove over a duplex groove. It stabilizes both the TAT triplex and mixed-base DNA triplexes better than known DNA minor groove binders (which usually destabilize the DNA triplex) and polyamines. Intercalator–neomycin conjugates were shown to simultaneously probe the base stacking and groove surface in triplex DNA. The groove recognition of aminoglycosides was not limited to DNA triplexes, but also was extended to RNA and hybrid triple helical structures.

3.2.4 Small Molecules that Interact with G-quadruplexes

Significant levels of telomerase activity have been detected in 85% of tumors; telomerase thus presents a target with potentially good selectivity for tumors over healthy tissue, and telomerase inhibition has been proposed as a new approach to cancer therapy.[80–83] One approach toward achieving the inhibition of telomerase is targeting the telomere. By considering the unique nucleic acid secondary structures associated with the telomerase reaction cycle, the G-quadruplex formed by the folding of a single-stranded G-rich overhang produced by telomerase activity is one of the candidates. As described in Section 3.1.1.4, G-quadruplexes can be formed by one or several strands of DNA; they thus display a wide range of structural polymorphisms owing to the number and orientation of the strands, the number of stacked G-tetrads, and variations in the location of the loop and its length and type. Nevertheless, each G-quadruplex contains at least one G-tetrad core that consists of two stacked G-tetrads. Hence, there are several common sites in G-quadruplexes that can be stacked or attacked by small molecules, such as terminal G-tetrads, the middle G-tetrads, the grooves/loops/backbones, and the central channel.[32]

3.2.4.1   Small Molecules Stack on the Terminal G-tetrads of a G-quadruplex to Stabilize its Structure

End-stacking is a primary interaction mode of a small molecule with a G-quadruplex, because the terminal G-tetrads have a large planner structure, and are easily stacked by small molecules. Small molecules with a planar aromatic structure are likely to interact with the terminal G-tetrads through π–π stacking, which is mainly controlled by hydrophobic and van der Waals interactions. Hence, the initial G-quadruplex ligand design was primarily based on molecules with aromatic components. The molecule that was first reported and extensively studied was 5,10,15,20-tetra-(N-methyl-4-pyridyl)-porphyrine (TMPyP4), a planar molecule with four methylated nitrogen groups that improve its solubility in aqueous solution (Figure 3.6)[83,84] Based on the crystal structures of quadruplex DNA and TMPyP4, a model of the interaction of G-quadruplex with TMPyP4 molecules was built by molecular modeling.[83] Porphyrins were stacked above and below the G-tetrads. The model showed that TMPyP4 is a good fit for stacking with G-tetrads, where it can be oriented to place each of the cationic N-methylpyridine groups into each of the four grooves of the quadruplex.[83] The interaction model later was confirmed by NMR studies with the structure of a 24-nt five-guanine tract sequence from the guanine-rich strand of the *MYC* NHE III$_1$ in K$^+$ solution.[85] The G-quadruplex–TMPyP4 complex gave well-resolved NMR spectra suitable for structure determination, which unambiguously suggested that TMPyP4 was stacked over the top G-tetrad. The positive charges of the ligand are in close contact with the negative phosphates of G2, A3, G5, and G12, as shown in Figure 3.7. The result also emphasized the stacking and the electrostatic contributions to the stabilization of a G-quadruplex.[85]

Many other small molecules are reported to stack with G-quadruplexes and have recently been reviewed in detail.[32] Similarly, metal complexes with planar ligands also like to bind to G-quaduplexes *via* end-stacking.[86–89] A typical example is Mn^{III} porphyrin complexes, which exhibited both high affinity and excellent selectivity for G-quadruplexes.[90] This is possibly contributed by the metal ions, which are necessary for G-quadruplex stabilization.

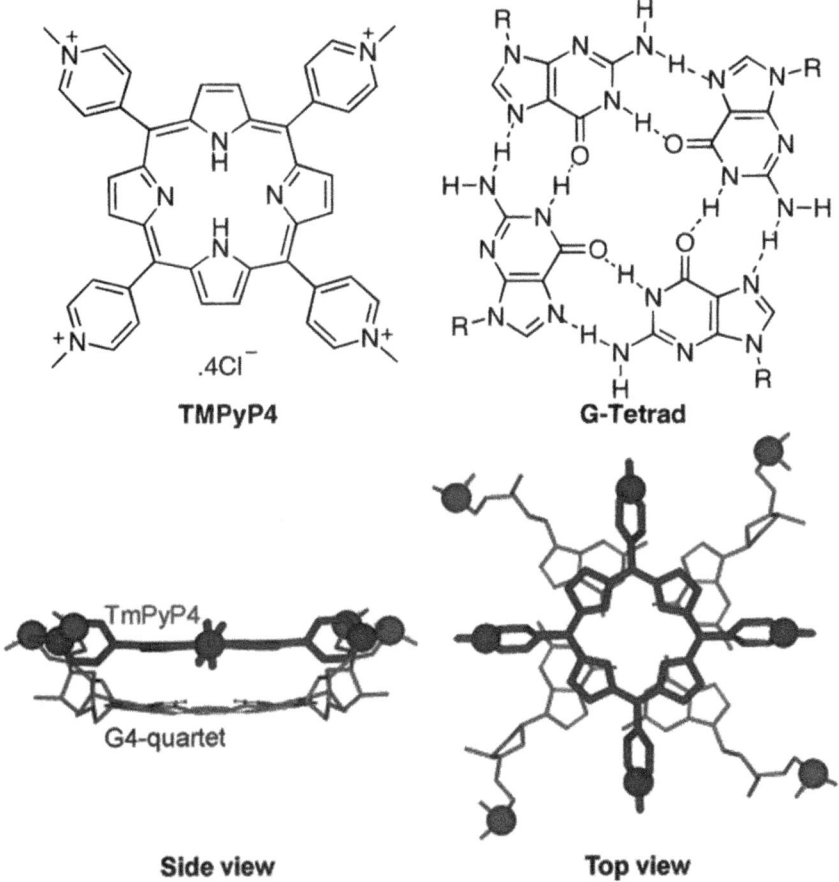

TMPyP4

G-Tetrad

TmPyP4

G4-quartet

Side view   **Top view**

Figure 3.6 Orientation of *N*-methylpyridinium ions of TmPyP4 with respect to phosphate ions of the G4 quartet. Thick and thin lines represent TmPyP4 and G4 quartet, respectively.[84] Reprinted with permission from ref. 84. Copyright 2006 American Chemical Society.

3.2.4.2   Small Molecules Intercalate into G-quadruplexes

Motivated by the layered structural feature of G-quadruplexes, intercalating into G-quadruplexes is one possible interaction mode for small molecules just like B-DNA intercalators, which have been extensively studied and reviewed.[91–93] Molecules with a planar structure can certainly insert into stacked G-tetrads well, as in B-DNA. However, possibly due to the distance between the layers of the G-tetrads being small, and there are metal ions in the center of the G-tetrads, very few G-quadruplex intercalators have been reported. Using a G-quadruplex with the general sequence $C_4T_4G_4T_{1-4}G_4$, Anantha *et al.* showed that H_2TMPyP binds to the quadruplex formed by T_4G_4 or the duplex formed form $(CG)_2ATAT(CG)_2$ *via* intercalation with high affinity at low [porphyrin]/[DNA] ratios based on the visible and CD spectra

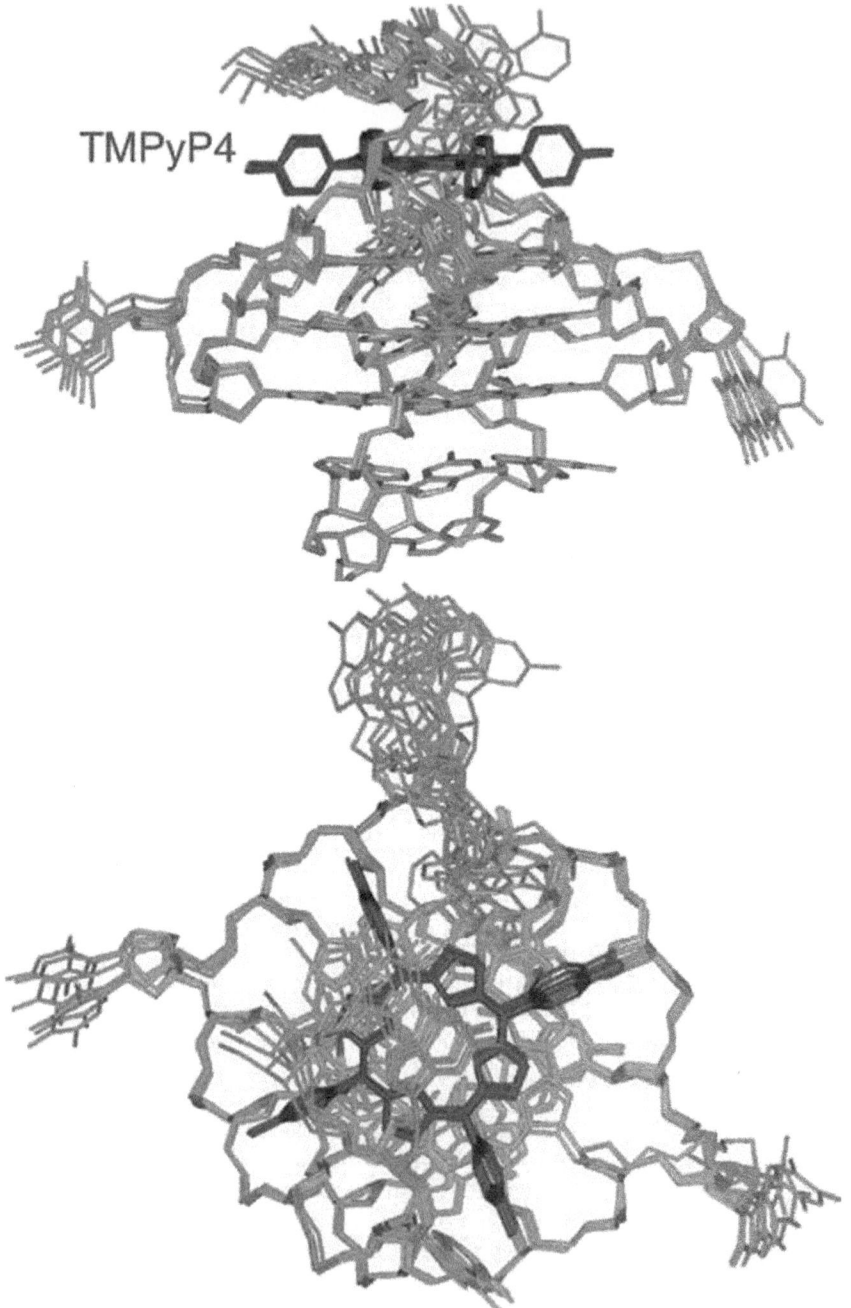

TMPyP4

Figure 3.7 Six superposed refined structures of the Pu24I quadruplex-TMPyP4 complex. (a) Side view. (b) Top view. Pu24I quadruplex is colored light gray; TMPyP4 is colored dark gray.[85] Reprinted with permission from ref. 85. Copyright 2005 Nature Publishing Group.

as well as the energy transfer data.[94] Further, they showed that H_2TMPyP bound to the quadruplex two times more strongly than it did to the duplex.[94] Porphyrins binding to the tetraplexes were also examined by isothermal titration calorimetry (ITC) and spectrophotometry under conditions that favor self-assembly to its intermolecular or intramolecular tetraplex structures. Analysis of the ITC and optical data revealed saturating porphyrin/tetraplex binding stoichiometries of 1:1, 2:1, and 3:1 for tetraplexes with different G contents, $d(G_2T_2G_2TGTG_2T_2G_2)$, $d(AG_3[T_2AG_3]_3)$, and $[d(T_4G_4)]_4$, respectively, involving near-equivalent sites.[95] Importantly, the stoichiometries correspond to the number of intervals between successive G-tetrad planes in each tetraplex. The results indicated binding by threading intercalation at each closely similar GpG site, rather than through either external electrostatic processes or end-pasted stacking modes. This mechanism was further supported by dynamic molecular modeling simulations with two DNA tetraplexes, which showed that stable intercalated complexes can be realized (Figure 3.8).[95] However, in these examples, no crystal structure or NMR data of the quadruplex–small molecule complexes were reported to support the proposal. Though G-quadruplexes contain large aromatic layers that are much bigger than in duplex B-DNA, small molecules with high affinity and high selectivity have not been found, which may need to be explored further.

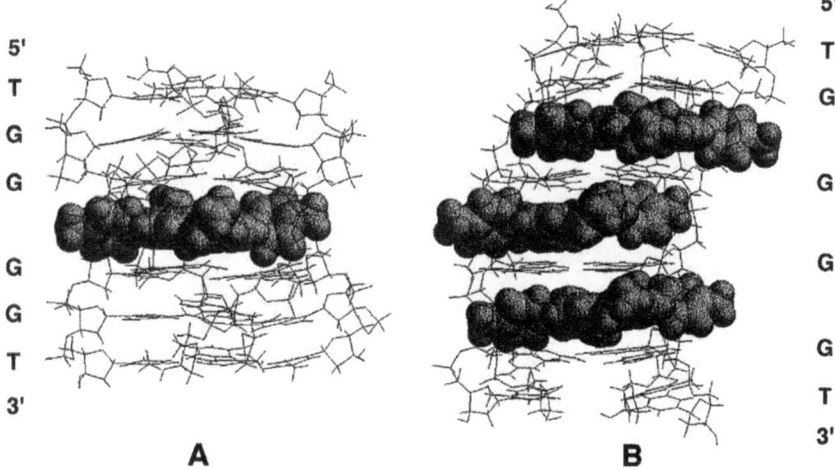

Figure 3.8 Energy-minimized structures for possible intercalation of the $[d(TG_4T)]_4$ tetraplex with (A) 5'-TGG*GGT (1:1 binding) and (B) 5'-TG*G*G*GT (3:1 binding), where each asterisk denotes a H_2TMPyP ligand. The complexes are viewed looking toward the G-tetrad stack with the porphyrin(s) shown in space-fill mode. Note the axial offset of the porphyrin relative to the long helical axis in the 1:1 core-intercalated complex and the alternating pattern of ligand displacements in the 3:1 complex.[95] Reprinted with permission from ref. 95. Copyright 1999 American Chemical Society.

3.2.4.3   Small Molecules Bind to the Grooves/Loops/Backbone of a G-Quadruplex to Stabilize its Structure

As described in the last section, there are significant size and shape distinctions between the grooves of G-quadruplexes and duplex B-DNA. In addition, the polymorphism of G-quadruplexes results in many G-quadruplexes with unique structural features, which may lead to some specific small-molecule ligands for G-quadruplexes. The ligands that bind to the groove of G-quadruplexes are similar to those that bind to B-DNA whereby the negative backbone attracts the positive small molecules matching the shape of the groove through electrostatic interaction or hydrogen bonds. For instance, Distamycin A is a small molecule with antibiotic properties, which binds with high affinity to B-DNA, and NMR studies indicated that distamycin molecules also could interact with [d(TGGGGT)]$_4$ tetraplexin in a 4:1 binding mode, with two distamycin dimers simultaneously binding two opposite grooves of the quadruplex as shown in Figure 3.9.[96] Since groove dimensions of G-quadruplexes vary according to the type of quadruplex, groove binding could also offer the opportunity to obtain molecules with high selectivity for a particular quadruplex structure. Similarly, these interactions could also occur between the loops/backbone of G-quadruplexes and small molecules

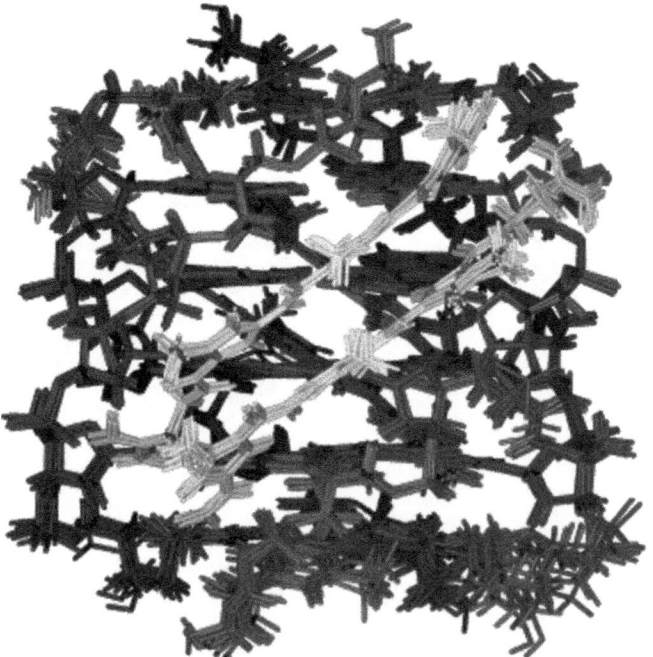

Figure 3.9   Side view of the superimposition of the 10 best structures of the 4:1 complex Dist-A/[d(TGGGGT)]$_4$. Dist-A is reported in light gray and DNA is colored in dark gray.[197] Reprinted with permission from ref. 197. Copyright 2007 American Chemical Society.

through electrostatic and hydrogen bonding.[97] Recently, it was also found that small molecules could also bind to G-quadruplexes through multiple binding sites at the same time.[98]

3.2.5 Small Molecules Induce and Promote the Stability of i-Motif Structure

The biological role of i-motif formation is currently uncertain because it has not been found *in vivo*.[99] The concern with the occurrence of i-motif DNA is its marginal stability at neutral pH. However, it is possible that under physiological condition, the pH of its environment can be modulated by its binding with macro-biomolecules or small molecules. Therefore, there is considerable interest in the design of small molecules to induce i-motif formation and promote its stability. Compared to the wealth of knowledge on the design of small molecules that interact with G-quadruplexes based on their structural features, the i-motif literature is comparatively scant with very few well-documented examples of interacting molecules. There are two categories of small molecules that could bind to i-motif DNA: organic molecules and metal complexes, as with G-quadruplexes. However, compared to G-quadruplexes, most of the ligands that bind to i-motif DNA are not specific, possibly because i-motif DNA only contains two bases intercalating alternatively, and the planar structure formed by two bases is rather small. Generally, the small molecules bind to i-motif DNA *via* stacking at the ends of the i-motif, groove binding, or intercalating, as with G-quadruplexes.

Fedoroff *et al.* showed in 2000 that cationic porphyrin TMPyP4, a quadruplex ligand, could also promote the formation of the i-motif DNA structure.[39] They investigated the binding properties of the cationic porphyrin TMPyP4 (structure in Figure 3.6) to human telomeric sequences d(CCCAAT)$_4$ and d(AATCCC)$_4$, and found that TMPyP4 bound and promoted the formation of i-motif DNA at pH 4.5. The affinity for i-motif DNA (dissociation constant, 45 µM) is weaker than its affinity for G-quadruplexes (dissociation constant, 0.5 µM) and duplex structures (dissociation constant, 1.2 µM). However, there was no significant change in the i-motif melting temperature on addition of the ligand ($\Delta T_m < 20$ °C). The interaction mode of TMPyP4 with i-motif DNA was proposed based on modeling with the NMR structure.[39] A model for the TMPyP4-(dAACCCC)$_4$ complex at a 2 : 1 ligand : tetraplex ratio is shown in Figure 3.10. This structure has been derived from restrained molecular dynamics and energy minimization simulations. Two porphyrin molecules were bound at opposite sides of the tetraplex in a symmetrical orientation (Figure 3.10). The binding geometry is close to the "face-on" binding model proposed for porphyrin binding to double stranded B-DNA. TMPyP4 binding to i-motif DNA has also been studied with the intramolecular human telomeric i-motif and a mutant sequence where the adenines had been substituted by thymines to alter the loop interactions.[100] In this case, it was found that loop interactions were important in the binding mode as binding properties were different for the native and the mutant sequences. Furthermore,

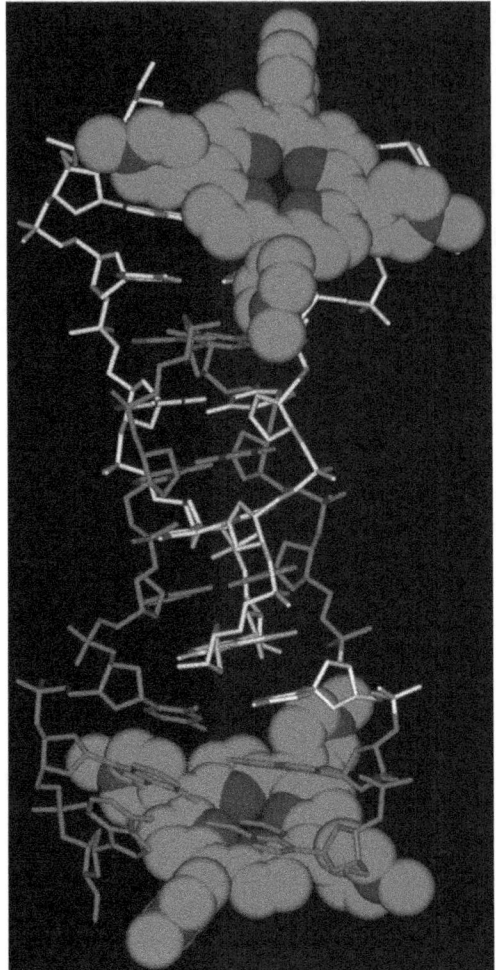

Figure 3.10   NMR-based model of the 2 : 1 TMPyP4-(dAACCCC)$_4$ complex. One C·C$^+$
parallel duplex is colored light gray and the other dark gray. The C3
residues of the two chains are white. TMPyP4 molecules are shown
in the CPK representation. All hydrogen atoms have been omitted.
The intermolecular distance restraints for C3H1′-*meta*-pyridyl H,
C3H4′-*meta*-pyridyl H, C3H4′-*ortho*-pyridyl H, A1H1′-*meta*-pyridyl H,
and A1H4′-*meta*-pyridyl H proton pairs have been applied during
molecular modeling.[39] Reprinted with permission from ref. 39. Copyright 2000 American Chemical Society.

the binding affinity is higher than that reported by Hurley and co-workers,
suggesting the different binding modes.[39] These results suggested that
TMPyP4 can trigger the formation of unusual DNA structures in both strands
of the telomeres, which may in turn explain why TMPyP4 has already been
shown to down regulate the expression of c-MYC through interaction with

the G-quadruplex, as well as inhibiting telomerase and tumor growth.[101] However, these interaction modes with i-motif DNA were not specific to i-motif DNA. Since the structure of an i-motif–small molecule complex has not yet been determined, their interaction is still questionable.

Apart from TMPyP4 and its derivatives, several other types of molecules could also interact with i-motif DNA to stabilize the structure or induce the formation of i-motif DNA.[99,102] However, most of them interact with i-motif DNA through the end-stacking mode; other modes are barely reported. In the case of i-motif DNA, intercalation binding is even scarcer than with G-quadruplexes because the size of the planar layer is small and alternating, and π–π stacking interaction is less possible. In addition, the distance between the base pairs layer is about 3.1 Å, which is smaller than B-DNA. Much more work is required, however, to evaluate this possibility.

3.3 Novel DNA Modulation Systems

3.3.1 Carbon Nanotubes (CNTs)

Carbon nanotubes (CNTs), discovered in 1991 by Kroto and Iijima, are made of rolled-up graphene sheets with one-dimensional extended π-conjugated structures. Each carbon atom is joined to three neighbors and the bonding in carbon nanotubes is sp^2.[103] CNTs are classified into three types according to the number of graphene layers: single-walled carbon nanotubes (SWNTs), double-walled carbon nanotubes (DWNTs), and multi-walled carbon nanotubes (MWNTs), which have one, two, and three or more walls, respectively (Figure 3.11).[104,105] Due to their unique structural, chemical and physical properties, CNTs have been extensively applied in nanoelectronics,[106,107] energy storage,[108] composites,[109] biotechnology,[110-112] medicinal chemistry, *etc.*[113,114] The combination of CNTs with biologically important structures, such as DNA or proteins, is particularly intriguing since it opens the door to novel biology and nanotechnology applications. Here, the interaction of CNTs with DNA including normal DNA and non-canonical DNA is summarized.

3.3.1.1 Interaction of CNTs with DNA

Given the remarkable properties of the DNA molecule and the unique properties of CNTs, the study of the interactions between CNTs and DNA has been an active research area. The study was initiated by the outstanding findings of the DNA assisted dissolution of SWNT in 2003.[115] Ultrasonication facilitates the separation of aggregated CNTs, producing individually solubilized nanotubes wrapped with DNA (CNT–DNA). The aromatic nucleotides of DNA interact with the hydrophobic CNT surface *via* π-stacking, while the polyanionic DNA backbone confers water solubility, thus leading to dissolution of CNTs. The strength of the binding interactions between CNTs and single-strand DNA (ssDNA) was demonstrated by the ability of the DNA to disrupt

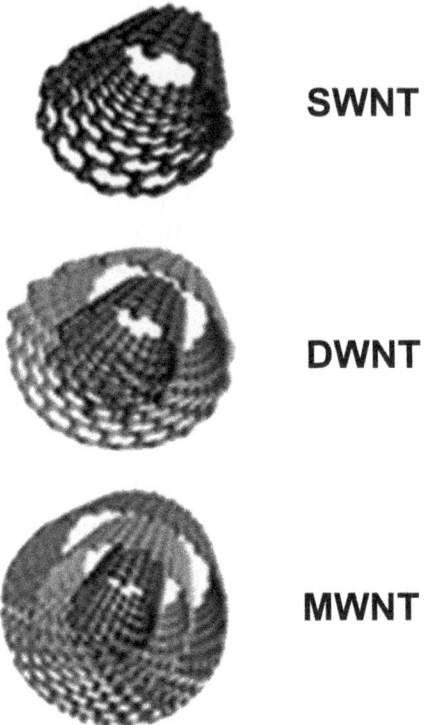

SWNT

DWNT

MWNT

Figure 3.11   Structures of SWNTs, DWNTs, and MWNTs.[105] Reprinted with permission from ref. 105. Copyright 2013 American Chemical Society.

the strong intertube interactions responsible for CNT aggregation into bundles. Individual dissolution of SWNTs using double-strand DNA (dsDNA) and ssDNA has been reported by different groups.[116] Generally, it is believed that DNA can be covalently or noncovalently functionalized with CNTs. Based on analysis of the vibration modes of surface enhanced IR absorption spectra, it was found that DNA had conformational changes after binding to SWNT, and three binding modes were proposed.[117] SWNTs wrapped helically by ssDNA have also been observed by means of the AFM technique.[118] Measurements and modeling of the regular AFM pattern observed along CNT–DNA suggest that the hybrids are composed of oligonucleotides closely arranged end-to-end in a single layer along the entire nanotube surface, with each turn of the wrapped DNA generating one surface peak in the AFM images. No significant impact of the oligonucleotide length was observed on the regular pitch of the surface pattern or on the width of the peaks along the CNT.

Meng *et al.* proposed a scheme to describe the interaction between nucleotides and CNTs based on measurement of electronic features through a local probe such as scanning tunneling microscopy (STM), and they demonstrated through quantum mechanical calculations that these

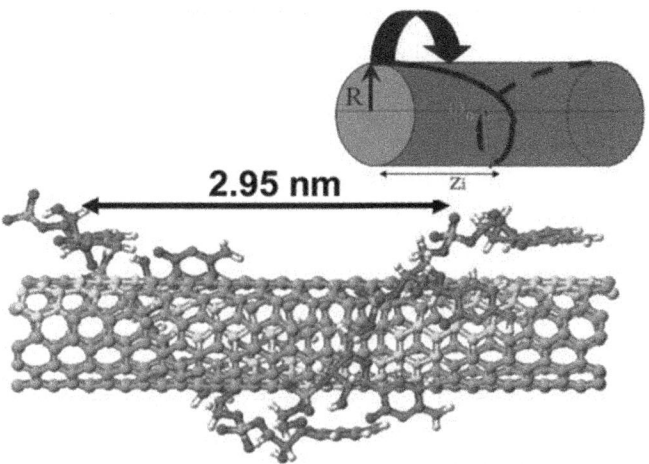

Figure 3.12 Schematic representation of DNA wrapping around a CNT (inset) and an initial configuration of the C-oligomer wrapped along the (6,5) tube chirality. Carbon atoms of the CNT marked by light gray indicate the direction of DNA wrapping with seven cytosine bases per helical turn, which lay parallel to the tube surface and nearly normal to the tube chiral vector, thus increasing the π–π overlap between the base and tube orbitals.[120] Reprinted with permission from ref. 120. Copyright 2009 American Chemical Society.

measurements can achieve 100% efficiency in identifying DNA bases.[119] STM studies showed that strands of DNA wrap around nanotubes at ~63° angle with a coiling period of 3.3 nm. This is the first topographic image of DNA–CNT hybrids with significant morphological detail (Figure 3.12).[120] The finding is in agreement with the theoretical predictions based on true *ab initio* and density functional theory (DFT) methods, as well as molecular orbital (MO), tight-binding (TB), and classical molecular dynamics (MD) simulations.[121] The ssDNA molecules interact more effectively with the surface of the CNT through a parallel orientation (quite flexible in bond torsion within the sugar–phosphate backbone). To extract more quantitative information about the observed DNA wrapping geometry, force field MD and first principle quantum mechanical calculations were employed, and the best π-stacking between a CNT and DNA was believed to occur when the orbitals of a DNA base align with the direction of the maximum MO density of a CNT, and the distance and binding energy of DNA to CNT were obtained. The thermodynamic stability of the DNA–SWNT complex is good; it can remain intact for at least 1 month. The dissolution of ssDNA from CNTs is believed to be highly sequence dependent and poly-(dT) and d(GT) provided the highest concentration of individual SWNTs aqueous solutions, but the dissolution of dsDNA is not fully understood. Short dsDNA shows higher efficiency of dissolution than that of genomic long dsDNA.[104,122]

The open ended CNTs also provide internal cavities (1–2 nm in diameter) that are capable of accommodating biomolecules. It was also proposed that DNA could be encapsulated inside SWNTs in a water solute environment.[123,124] Gao *et al.* employed a classical molecular dynamics simulation for a SWNT–DNA conjugate in an aqueous environment, using an uncapped armchair (10,10) SWNT and a ssDNA oligonucleotide with 8 adenine bases. They found that ssDNA spontaneously encapsulated into (10,10) SWNT, mainly driven by van der Waals and hydrophobic interactions, with the former being dominant. The diameter (1.08 nm) of (8,8) was suggested as an onset for the encapsulation of ssDNA.[123] The DNA molecules can be encapsulated, provided that the tube size exceeds a certain critical value, which was also evidenced experimentally. The interaction strength of SWNT–DNA conjugates may also depend on the diameters[125] and the electronic properties (*i.e.*, metallic or semiconducting)[126–128] of the SWNT and the types of bases in the DNA.[129–131]

3.3.1.2 SWNTs Induce and Stabilize i-Motif Structure

B–Z DNA transition has been achieved on the surfaces of SWNTs through binding to the DNA major groove.[132,133] Qu and his co-workers reported that SWNTs can selectively stabilize i-motif DNA but not G-quadruplex DNA.[40] Through UV melting, NMR, S1 nuclease cleavage, CD, and competitive FRET methods they showed that SWNTs could induce i-motif formation by binding to the 5′-end major groove under physiological conditions or even at pH 8.0, and this could compete with DNA duplex association. It was proposed that charge stabilization due to interactions between the positively charged CC^+ base pairs and SWNTs lowered the pK_a of the CC^+ base pairs and induced i-motif formation. The proposal was supported by the chemical modification of SWNT into $SWNT-COO^-$ and $SWNT-CONH-CH_2CH_2-NH_2^+$; the latter decreased the T_m of the i-motif, suggesting that a negatively charged carboxyl group on the open end of the side wall of $SWNT-COO^-$ enhanced interactions with i-motif CC^+ base pairs by providing favorable electrostatic attractions. However, the proposal that SWNTs interact with the positively charged CC^+ base pairs to lower the pK_a of the CC^+ base pairs may need to be confirmed by the structure of the SWNTs/i-motif complex.

3.3.2 Graphene Oxide (GO)

Graphene, a single atomic layer of carbon atoms that are chemically bonded with a hexagonal symmetry, shows a true two-dimensional crystalline structure and unique properties, and has stimulated extensive applications in a variety of fields since it first became available in the laboratory.[134–139] It is similar to an opened SWNT. However, similar to the CNTs and other carbon materials, the lack of surface functionalities and the poor dispersion capabilities of graphene sheets in aqueous solution and most organic solvents make the processing of the graphene sheets extremely hard, and severely limit their applications, especially for use in biological systems.[140] As shown in Figure 3.13, graphene

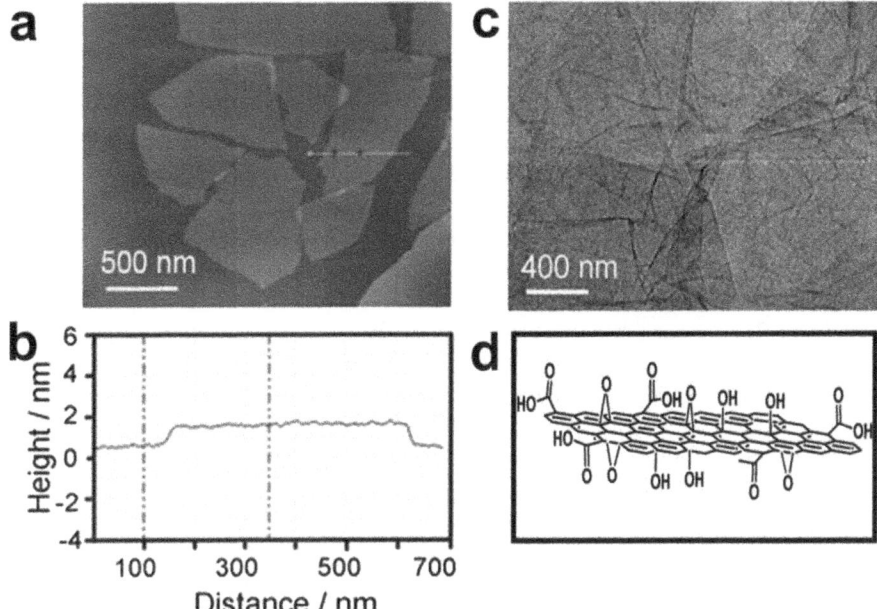

Figure 3.13 (a) A tapping mode AFM image of graphene oxide (GO) sheets on a mica surface, (b) the height profile of the AFM image, (c) TEM image of GO, and (d) chemical structure of GO.[144] Reprinted with permission from ref. 144. Copyright 2010 American Chemical Society.

oxide (GO) sheets maintain all structural features of graphene, but have abundant surface oxygen-containing groups, including epoxide, hydroxyl, and carboxylic groups, hence GO sheets show good solubility in aqueous solution and some polar organic solvents, and can be processed easily through a wet chemical procedure.[134] Additionally, the GO sheets could be prepared through the oxidative intercalation and exfoliation of graphite on a mass scale. The surface functional groups can also provide plenty of reaction sites for linking external species, such as small molecules or macromolecules, without using any cross-linking reagents or additional surface modification.[141-148] More importantly, the large and atomically flat structural feature of GO provides a platform for studying its interaction with external species through high resolution surface analytic techniques, such as atomic force microscopy.

3.3.2.1 Interaction of GO with Single-strand DNA (ssDNA)

It was first reported in 2009 by He *et al.* that GO could react with ssDNA through π-stacking interaction and hydrophobic interaction between the ring structures in the nucleobases and the hexagonal cells of the graphene.[149] Anisotropy, fluorescence, NMR, and CD studies suggested that ssDNA was promptly adsorbed onto functionalized graphene forming strong molecular interactions that prevent DNase I from approaching the constrained DNA.[150]

GO adsorbs ssDNA more quickly than dsDNA, allowing for selective removal of the former with a simple mixing and centrifugation operation.[151] Many DNA detection systems based on this fact have been reported.[139,152-154] For instance, a system with an acridinium ester (AE)–hydrogen peroxide and GO was established, in which both probe DNA and AE were absorbed on the surface of GO in the absence of target DNA, producing a weak CL emission owing to the CL quenching effect of GO. In the presence of target DNA, dsDNA was generated leading to the release of the oligonucleotide from the GO surface. Thus, the quenching effect of GO will be no longer effective and a strong CL signal can be observed.[155,156]

Various fluorescent biosensors have been successfully developed for the detection of other molecules, such as proteins, small molecules, and metal ions, using the interaction of GO with DNA. In these designs, the high fluorescence quenching efficiency of GO as well as its high differentiation ability for different structures of bioprobes were employed to achieve simple, sensitive, and specific detection. However, the detailed mechanism of the interaction between GO and DNA is not well understood. Recently, Zhang *et al.* investigated the size effect of GO on the interaction between DNA and GO based on the fluorescence quenching ability of GO.[157] Three GO samples with different sizes and fluorescence labeled DNA were employed. It was found that the quenching efficiency of GO1 (~500 nm) was ~3.6 times higher than that of GO3 (~40 nm). The higher quenching efficiency probably originated from the relatively strong interaction between DNA and GO1, and the relatively weak interaction between small size GO and DNA. The results demonstrated that the GO–DNA interaction is size dependent. Even though both GO and DNA are negatively charged, the DNA can still be adsorbed on the surface of GO.[158] The forces that drive the adsorption include π–π stacking interaction and hydrophobic interaction.[158,159] The negative charges on the surface of GO contribute much less when its size is large (>200 nm). However, when the size of GO is decreased to 40 nm, the charge density increases because of the repeated oxidation, and the negative charges provide a reverse force that drives the DNA apart from GO, which leads to lower quenching efficiency. The sp^2 domains in the basal plane of GO and reduced GO may provide the main driving force for the binding of DNA to them.[158] However, this only can partially explain the results, because they also found that the quenching efficiency was inversely related to the length of DNA, suggesting that longer ssDNA interacts more weakly with GO, which needs to be further explored.

3.3.2.2   Interaction of GO with Double-Strand DNA (dsDNA)

As described in Section 3.3.2.1, ssDNA is believed to be absorbed on the GO surface mainly through π–π stacking of bases on the GO, so it was generally believed that interaction of dsDNA with GO is weak. However, Lei *et al.* found that dsDNA can bind to GO forming complexes (dsDNA/GO) in the presence of certain salts (Figure 3.14).[160] On the other hand, they found that a nonionic surfactant, such as triton X-100, can block the formation of dsDNA/GO complexes, suggesting that the origin of the affinity of dsDNA to GO might be the π–π interaction between the aromatic region of GO and nucleobases that are exposed due to the breathing fluctuations of the dsDNA. This interaction has been used to show that GO can protect dsDNA from enzymatic digestion through hindering the access of DNA enzymes to dsDNA. Colloidal chemically converted graphene sheets (CCGs) are able to contact with each other to form stable self-assembled stacked-CCGs *via* salt "invasion", and could also capture dsDNA (Figure 3.15).[161] It was recently reported that GO-adsorbed DNA was facilitated following its hybridization with cDNA on the GO surface; when the GO surface was almost saturated with the adsorbed DNA, nonspecific desorption dominated the process through a simple displacement of the GO-adsorbed DNA molecules by the incoming DNA molecules because of the law of mass action.[162]

However, even with metal ions, the absorption of dsDNA on the surface of GO is still not fully understood. We found that GO can intercalate between

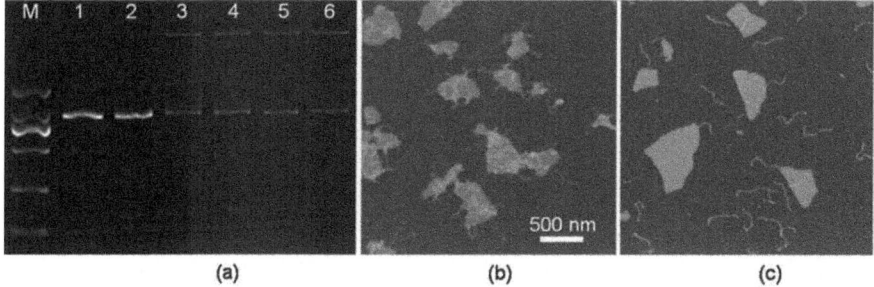

Figure 3.14 (a) An image of agarose gel electrophoresis of dsDNA and mixtures of dsDNA–GO in various buffer conditions. Lane 1: dsDNA only; lanes 2–6: mixture of dsDNA and GO (0.1 mg mL^{-1}) in water (Lane 2), or aqueous solutions of 2.5 mM MgCl$_2$ (lane 3), 5 mM CaCl$_2$ (lane 4), 250 mM NaCl (lane 5), 50 mM KCl (lane 6), respectively. (b and c) AFM images of a mixture of dsDNA and GO (0.15 mg mL^{-1}) prepared with DNase I reaction buffer (b), and with a pure water solution (c).[160] Reprinted with permission from ref. 160. Copyright 2011 The Royal Society of Chemistry.

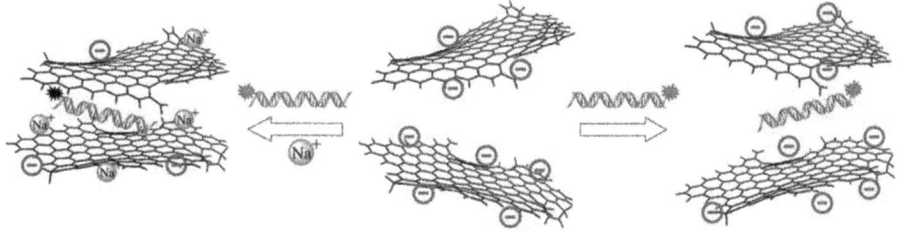

Figure 3.15 Schematic illustration of the salt-controlled assembly of stacked-CCGs for capturing dsDNA in aqueous solution.[161] Reprinted with permission from ref. 161. Copyright 2012 The Royal Society of Chemistry.

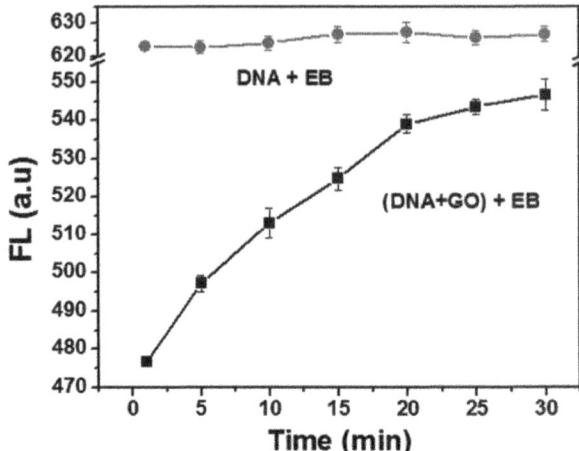

Figure 3.16   FL intensity variations of the samples of plasmid DNA with EB, and the DNA first incubated with GO (10 μg mL^{-1}), then with the same amount of EB (6 μg mL^{-1} DNA and 12 μg mL^{-1} EB).[163] Reprinted with permission from ref. 163. Copyright 2010 American Chemical Society.

base pairs of plasmid DNA based on gel electrophoresis and circular dichroism (CD).[163] The decrease in the intensity of supercoiled DNA and the appearance of the smeared bands of the gel in the presence of GO indicated that GO sheets interacted with the DNA and very likely in a nonspecific manner. The CD spectrum of DNA was dramatically changed upon addition of GO. The intensities of both the positive and negative ellipticity bands were decreased as the amount of GO increased, suggesting intercalative interaction between GO and dsDNA. This was also supported by the fluorescent experiment with EB, and it was found that the intercalated GO was gradually replaced by EB, resulting in the increase of FL when EB was present in the system (Figure 3.16). In addition, the helix–coil transition temperature (T_m) for the DNA increased by 16 °C in the presence of 3 g mL^{-1} GO.

3.3.3   Graphene Quantum Dots (GQDs) and their Interaction with i-Motif DNA

The large lateral sizes and inconsistency in the size of GO impede their applications in biological systems. Thus, graphene quantum dots (GQDs), which are graphene sheets with lateral dimensions less than 100 nm, have attracted much attention.[164,165] In comparison with graphene and GO with micrometer lateral dimensions, the quantum confinement and edge effects give GQDs many unique electronic and optoelectronic properties. The intrinsic characteristics of carbon materials (single atomic-layered structural motif, small lateral dimension, and abundant surface/edge groups) provide GQDs with decent inert chemical/biological stability. We have showed that in the GO/Cu^{2+} cleavage system, GQDs/Cu^{2+} is much more efficient, and GQDs exhibited better biocompatibility.[166]

As described in the section 3.2, small molecules with a large flat aromatic surface or a planar structure can bind to i-motifs or G-quadruplexes. The driving force for their interactions is the π–π interaction between the aromatic structure and the planar base-pair. SWNTs could selectively induce human telomeric i-motif formation by shifting the pK_a of the C:C$^+$ base pairs. As a matter of fact, GO and GQDs are very similar to these planar small molecules at a molecular level. We have previously demonstrated that GO sheets, which have single atomic layered structure, can intercalate with DNA plasmids.[163] These results triggered the research on the interaction of G-quadruplexes and i-motifs with GQDs. Unfortunately, we did not observe any interaction between G-quadruplexes with GQDs in terms of melting temperature and CD spectrum. In contrast, the interaction of GQDs with i-motifs was unexpectedly spotted. GQDs could stabilize i-motif DNA based on the increase in the melting temperature in the presence of GQDs. More interestingly, the GQDs can induce the formation of i-motifs under basic or neutral conditions (Figure 3.17). It is possible that GQDs intercalate to i-motifs as they do to double helical DNA. However, using the fact that the fluorescence emission of EB could be enhanced when it intercalates to DNA, it was found that GQDs interacted with i-motif DNA by end-stacking instead of intercalation (Figure 3.18). The end-stacking mode is rational, because the distance between two adjacent base pairs in i-motifs is smaller (3.1 Å *vs.* 3.4 Å in the helix). In addition, in the four-stranded intercalated i-motif structure, neighboring C:C$^+$ base pairs from intercalated duplexes are arranged in a way that no overlap exists between their six-membered aromatic ring systems.[167] The interaction mode of i-motifs with GQDs is different to with SWNT,[40,168,169] possibly because of their different structural motifs. GQDs are more like planar small aromatic molecules, while SWNTs, although similar at the molecular level, are structurally unfavorable for interaction with DNA structures, and the diameter and length are much large than i-motifs or other unconventional DNA structures.

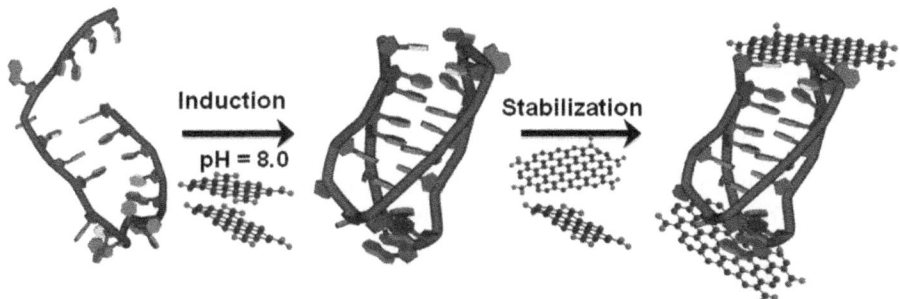

Figure 3.17 Proposed mechanism of the stabilization and induction of the i-motif structure by GQDs. Folded i-motif structure is generated from the PDB file (PDB id: 1EL2).[167] Reprinted with permission from ref. 167. Copyright 2012 American Chemical Society.

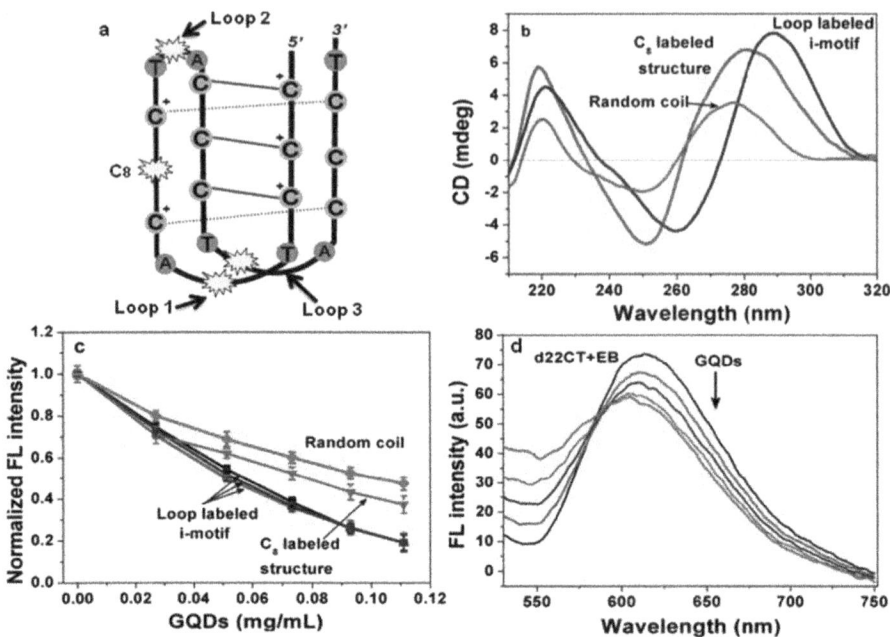

Figure 3.18   (a) Four 2-Ap-labeled sites in d22CT. (b) CD spectra of the 2-Ap-labeled samples. The random coil is the loop3-labeled sample at pH 8.0. (c) Fluorescence intensity changes of the 2-Ap-labeled d22CT structures in the presence of GQDs. (d) Interaction of the d22CT i-motif with ethidium bromide (EB) and GQDs.[167] Reprinted with permission from ref. 167. Copyright 2012 American Chemical Society.

3.4   Potential Applications of the Interactions of CNTs and GO with DNA

The small molecules that are potentially applicable as anticancer drugs and for gene manipulations have been well documented. In this section, the applications of CNTs and GO as gene carriers and anticancer drugs are reviewed.

3.4.1   Gene Delivery

3.4.1.1   CNTs as Gene Carriers

Due to their unique cylindrical structure and properties, CNTs can be used as carrier for genes. Pantarotto *et al.* showed for the first time that ammonium-functionalized CNTs (f-CNTs) are able to associate with plasmid DNA (pDNA) through electrostatic interactions.[170] These f-CNTs could penetrate into CHO cells membranes and exhibited low cytotoxicity, and the f-CNT-associated plasmid DNA can be delivered to cells efficiently. The gene expression levels of β-galactosidase (marker gene) were up to 10 times higher than those achieved with naked DNA alone (Figure 3.19). These

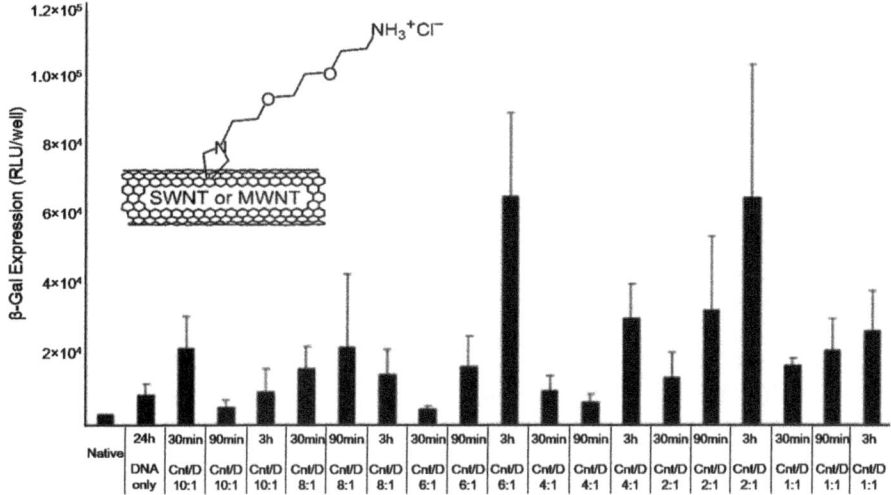

Figure 3.19   Delivery of plasmid DNA by f-SWNTs and expression in cells. Levels of marker gene (b-gal) expression in CHO cells in relative light units (RLU) per mg total protein.[170] Reprinted with permission from ref. 170. Copyright 2004 Wiley-VCH Verlag GmbH & Co. KGaA.

results indicated that the complex formed between f-CNTs and DNA can constitute a novel class of non-viral gene delivery system. However, the preliminary comparative gene expression data for the lipid:DNA and f-CNT:DNA delivery systems showed that the f-CNT was less effective for transfection *in vitro* than the lipid:DNA systems. Later studies showed that different types of cationic CNTs condensed DNA to varying degrees, indicating that both CNTs surface area and charge density were critical parameters that determine the interaction and the complex formation between f-CNTs with DNA.[171] For the same purpose, the CNTs were also modified with polyethyleneimine (PEI) groups, an overall highly cationic macromolecule with a high density of terminal amine groups that is commonly used as a transfection agent which.[172] Similarly, the CNT–PEI/pDNA systems produced much higher levels of gene expression than pDNA alone. The cellular penetration mechanism of the CNT–PEI/pDNA system was believed to be similar to PEI, because the transfection efficiency with the CNT–PEI constructs was about 3 times higher than with PEI alone. In contrast to covalent attachment of PEI to CNTs, non-covalent functionalization of SWNTs with PEI was also employed.[173] Two different routes for noncovalent attachment of PEIs with MWs of 25, 10 and 1.8 kDa to the side walls of CNTs were explored: (1) Conjugation of PEIs to phospholipid (DSPE) through a polyethylene glycol (PEG) linker; and (2) conjugation of PEIs to hydrocarbon chains of two different lengths forming amphiphiles. The non-covalent PEIs functionalized CNTs retained the ability to effectively bind and condense pDNA, and they were able to increase gene expression of pDNA by up to two or even three orders of magnitude greater than free pDNA, and several fold relative to the corresponding underivatized PEI bases. They also found that the largest increases in transfection activity relative to underivatized PEIs occurred with the shortest polycation tested and used the shortest alkyl chain tested for attachment to CNTs. Importantly, the authors moved the research a step forward, performed the gene transfer *in vivo* by intravenous injection into mice. However, *in vivo* studies clearly showed that the modified vectors in the CNT–PL–PEI group were not as effective as the vectors in the other two groups tested *in vitro*. Nevertheless, their findings highlighted the importance of *in vivo* testing for a reliable assessment of the transfection efficiency of a given vector.[173] Similar work was also reported by Gao *et al.*, in which they believed that although the MWNTs-based method was less efficient than commercial gene transfection agents, the MWNTs exhibited much lower toxicity.[174]

Liu *et al.* presented the synthesis of polyamidoamine-functionalized multiwalled carbon nanotubes (PAA-*g*-MWNTs) and their application as a novel gene delivery system.[175] The PAA-*g*-MWNTs showed comparable or even higher transfection efficiency than PAA and PEI at optimal w/w ratio. Intracellular trafficking of Cy3-labeled pGL-3 indicated that a large number of Cy3-labeled pGL-3 were attached to the nucleus membrane, the majority of which was localized in the nucleus after incubation with cells for 24 h (Figure 3.20).

Qin *et al.* also modified MWNTs with polyamidoamine (PAMAM) to increase the CNTs' gene loading, and eventually to improve the transfection performance of CNTs.[176] However, though the MWNT–PAMAM hybrid could deliver the GFP gene into cultured HeLa cells more efficiently than the unmodified CNTs, the efficiency was much lower than Lipofectamine 2000. The surfaces of SWNTs were also functionalized with cationic glycopolymers,[177] and the copolymer modified SWNTs were found to be biocompatible and exhibit transfection efficiencies that are comparable to the commercially available agent Lipofectamine 2000. Various modified CNTs were tested in different cell lines for gene delivery.[178–180]

In contrast to the above well-known strategies to modify CNTs for DNA delivery, Geyik *et al.* recently developed a new approach using the covalent attachment of linearized plasmid DNA to MWNTs.[181] Cifuentes-Rius *et al.* demonstrated that two different monomers, pentafluorophenyl methacrylate (PFM) and allylamine (AA), could be polymerized on the CNT surface in a home-built plasma reactor, allowing the formation of CNT-mediated gene delivery vectors.[182]

Except for the polymerization modification, CNTs were also fabricated with other materials, such as metals. A new strategy to deliver exogenous pDNA was also developed with CNT that contain Ni particles enclosed in their tips. Because of the presence of nickel, the complexes have some extra properties that can be used. It was shown that the gene expression was 80–100% of the cell

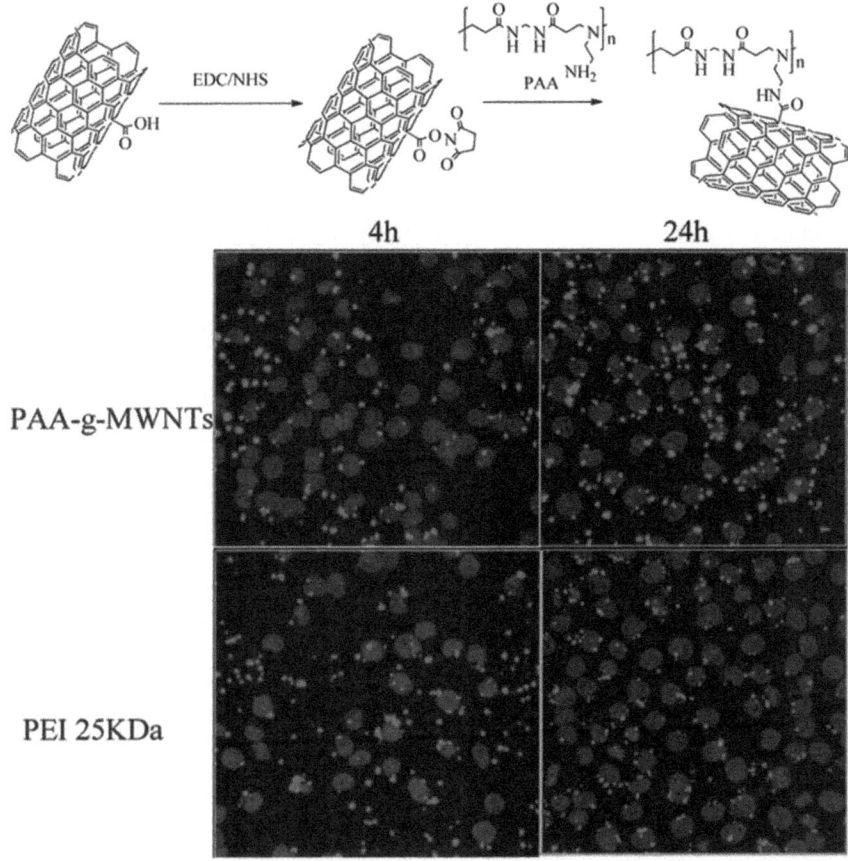

Figure 3.20 Intracellular trafficking of Cy3-labeled pGL-3 (light gray) when combined with PAA-*g*-MWNTs and PEI 25 KDa at optimal w/w ratio. The
localization of fluorescent particles in COS-7 cells was visualized at
4 and 24 h post-transfection. Nucleus (dark gray) was stained with
Hoechst 33258.[175] Reprinted with permission from ref. 175. Copyright
2011 American Chemical Society.

population in the presence of CNT with Ni, while CNT deprived of Ni particles
did not produce any gene expression. SWNTs were also dotted with Au nano-
crystals (Au–SWNTs) to bind DNA could and possibly act as a gene carrier.[183]

3.4.1.2 GO as Gene Carriers

The ability of GO to strongly bind to ssDNA yet without a strong interaction
with dsDNA molecules has been extensively applied in the gene delivery. Liu
et al. pioneered gene delivery by GO in 2011.[184] They employed GO bound
with cationic polymers, PEI with two different molecular weights at 1.2 kDa
and 10 kDa, forming GO–PEI-1.2k and GO–PEG-10k complexes as non-toxic

nano-vehicles for efficient gene transfection. These complexes were stable in physiological solutions and exhibited low cellular toxicity compared to the bare PEI-10k polymer. The positively charged GO–PEI complexes were able to further bind plasmid DNA (pDNA) for intracellular transfection of the enhanced green fluorescence protein (EGFP) gene in HeLa cells (Figure 3.21). They found that different molecular weight PEI actually exhibited different EGFP transfection efficiencies and toxicities. The results suggested GO to be a novel gene delivery nano-vector with low cytotoxicity and high transfection efficiency, promising for future applications in non-viral based gene therapy.

Shortly after that, Zhang *et al.* reported similar work.[185] Instead of using electrostatic attraction, GO were modified by branched polyethylenimine forming PEI–GO complexes *via* an amide bond with widely used EDC chemistry. As-prepared PEI–GO exhibited an excellent ability to condense DNA at a low mass ratio with a positive potential of 49 mV. A WST assay revealed that PEI–GO was significantly less cytotoxic than PEI 25 kDa. Interestedly, it was demonstrated that the luciferase expression of PEI–GO is comparable or even higher than that of the PEI 25 kDa at optimal mass ratio. Moreover, intracellular tracking of Cy3-labelled pGL-3 indicated that PEI–GO could effectively deliver plasmid DNA into cells and be localized in the nucleus. These findings further suggest that PEI–GO is a promising candidate for efficient gene delivery. However, with regard to DNA transfection efficiency, the PEI–GO is just comparable or slightly better than PEI 25 kDa under optimal conditions. Similar work was also published around that time by a Korean group.[186]

To further improve the performances of the these delivery systems, nuclear localized signals (NLS) peptide PKKKRKV (PV7, one of the primary NLS peptides) was introduced into GO–PEI (10 kDa)/DNA binary complexes to engineer a nuclei localized gene delivery system.[187] The *in vitro* transfection

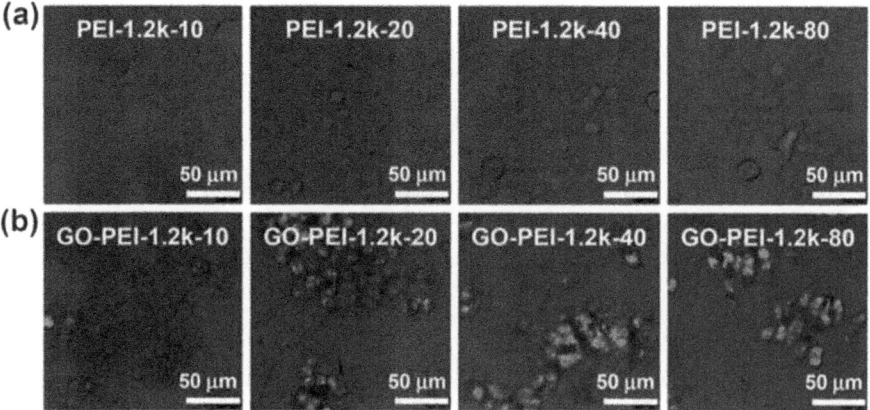

Figure 3.21   Confocal fluorescence images of EGFP transfected HeLa cells using PEI-1.2k (a), GO–PEI-1.2k (b), at varying N/P ratios from 10 to 80. Scale bar = 50 μm.[184] Reprinted with permission from ref. 184. Copyright 2011 The Royal Society of Chemistry.

experiments demonstrated that post-addition and simultaneous addition of PV7 would improve the gene expression efficiency, while the prior-addition of PV7 showed little change compared with GO–PEI/DNA binary complexes. However, the cytotoxicity of GO–PEI was much lower than PEI 10 kDa and PEI 25 kD against both HeLa cells and 293T cells. Therefore, the PV7 conjugated GO–PEI system compromised the contradiction between the cytotoxicity and the transfection efficiency, which could be an alternative strategy for a nuclear targeted gene delivery vehicle. Zhang *et al.* further improved the GO based gene delivery vector by coating GO with both polyethylene glycol (PEG) and PEI, obtaining a dual-polymer-functionalized nanoscale GO (nGO–PEG–PEI).[188] Employing this newly designed system as the pDNA transfection agent to treat Drosophila S2 cells that exhibit relatively low transfection efficiencies when treated by conventional nonviral agents offered about 7-fold and ~2.5-fold higher efficiency compared with those achieved by using bare PEI and Lipofectamine 2000, respectively. Interestingly, the advantages of nGO–PEG–PEI were even more dramatic when transfecting cells with lower-quality linearized DNA. When pDNA encoding EGFP (pTub-EGFP) was linearized to linear DNA by restriction digestion without damaging the EGFP gene, the transfection with linear DNA using nGO–PEG–PEI (N/P ratio = 30) was 25.36% ± 4.05%. Under the same conditions, neither PEI nor Lipofectamine 2000 were able to carry out efficient transfection. The authors thus believe that the high efficiency is contributed by their unique pathway of entering into insect cells, which may be different from the endocytosis of mammalian cells (Figure 3.22).

Based on these advances, a facile approach was developed to fabricate patterned substrates of nano-graphene oxide, demonstrating highly localized and efficient gene delivery to multiple cell lines in a substrate-mediated manner. In this system, GO substrates served as a valid platform to pre-concentrate PEI–pDNA complexes and maintain their gradual releasing for a relatively long period of time. Thus, it allowed successful gene delivery in selected groups of cells on the stripe-patterned GO substrates, without transfecting their neighbor cells directly cultured on glass. In addition, the authors also demonstrated that the GO substrates exhibited excellent biocompatibility and enabled effective gene transfection for stem cells, thus they may find application in stem cell research and tissue engineering.[189]

Recently, instead of using PEI in most work, Xu *et al.* developed a flexible two-step method to introduce the atom transfer radical polymerization (ATRP) initiation sites containing disulfide bonds onto GO surfaces.[190] Thus, surface-initiated ATRP of (2-dimethyl amino) ethyl methacrylate (DMAEMA) was then employed to tailor the GO surfaces, producing a series of organic–inorganic hybrids (termed SS–GPDs) for highly efficient gene delivery. Under reducible conditions, the PDMAEMA side chains can be readily cleavable from the GO backbones, benefiting the resultant gene delivery process. They also used the system for anticancer drug delivery.

In fact, other materials such as chitosan (CS) have also been used to modify GO. CS-grafted GO (GO–CS) sheets were able to condense plasmid DNA

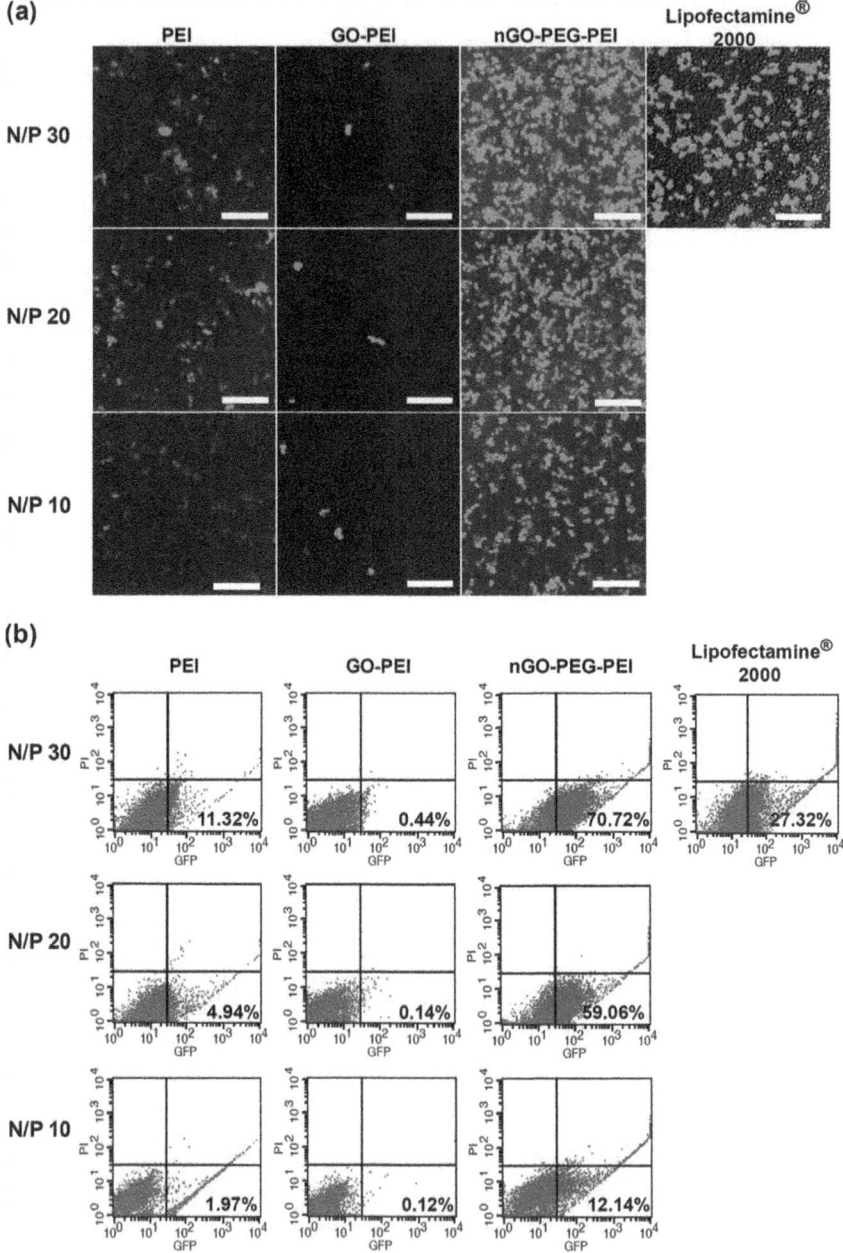

Figure 3.22   EGFP pDNA transfection for S2 cells. (a) Confocal laser scanning fluorescence images and (b) flow cytometric data of S2 cells measured
24 h post transfection using nGO–PEG–PEI, nGO–PEI and bare PEI
as transfection agents at different N/P ratios of 10:1, 20:1 and 30:1.
Lipofectamine 2000 was used as a control, according to the standard
protocol recommended by the vendor. Freshly prepared pDNA encoding EGFP with a dose at 1 µg was used in the above experiments. Scale
bar = 100 µm.[188] Reprinted with permission from ref. 188. Copyright
2013 Wiley-VCH Verlag GmbH & Co. KGaA.

into stable, nanosized complexes, and the resulting GO–CS/pDNA nanoparticles exhibit reasonable transfection efficiency in HeLa cells at certain nitrogen/phosphate ratios, as shown in Figure 3.23.[191] Latter, chitosan derivatives in the form of folate conjugated trimethyl chitosan (FTMC)/graphene oxide (GO) nanocomplexes (FG-NCs) prepared *via* electrostatic self-assembly were also developed as a targeted delivery vehicle for both doxorubicin (DOX) and plasmid DNA.[192] Based on a similar concept, GO was also functionalized by conjugating with PEG and folic acid, followed by the loading of siRNA with the aid of 1-pyrenemethylamine hydrochloride *via* π–π stacking for the

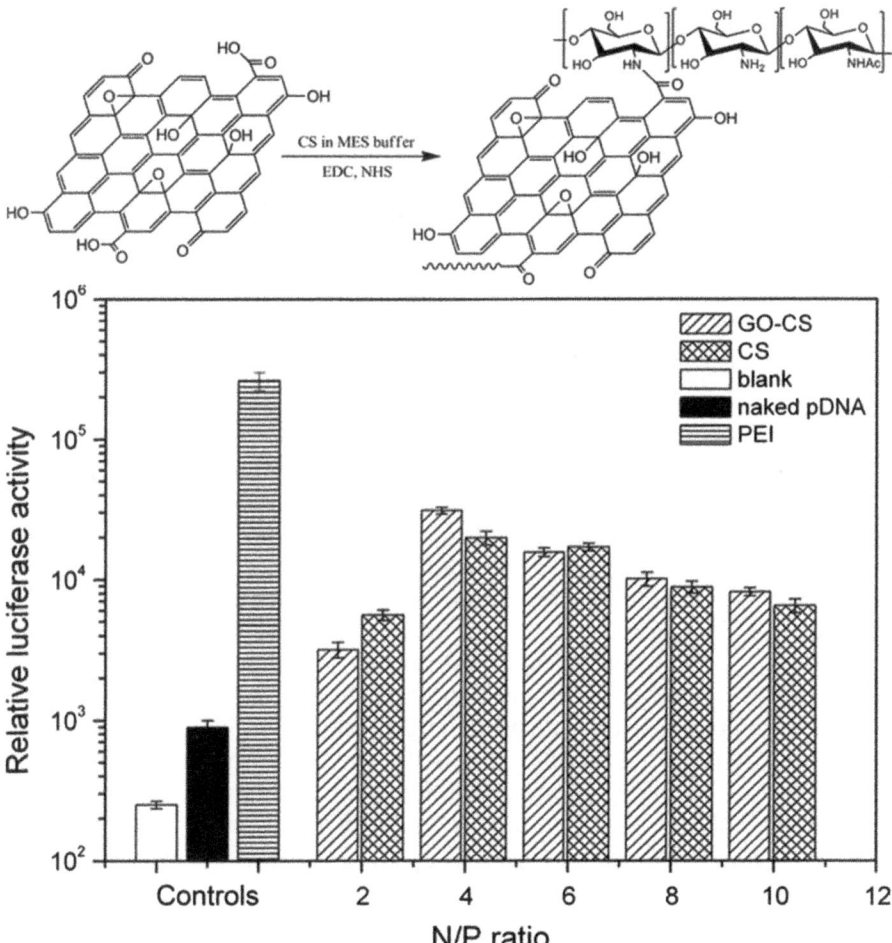

Figure 3.23 *In vitro* gene transfection efficiency of GO–CS/pDNA complexes in comparison with that mediated by pristine CS at different N/P ratios in HeLa cells in presence of serum after 24 h. Naked pDNA (ND) and 25 kDa PEI at its optimal N/P = 10 were used as positive controls. Data represent mean ± standard deviation ($n = 3$).[191] Reprinted with permission from ref. 191. Copyright 2011 Wiley-VCH Verlag GmbH & Co. KGaA.

targeted intracellular delivery of hTERT siRNA. GO–PEG–FA–PyNH2 specifically transfected the FAM-labeled DNA to the HeLa cells. RT-PCR and West-
ern blot assays showed that hTERT siRNA delivered *via* GO–PEG–FA–PyNH2
silenced the mTERT expression and suppressed the expression of mTERT
protein in HeLa cells.[193]

3.4.2 Anticancer Drugs

Due to their unique cylindrical structure and properties, nanotubes are used
as carrier for genes, and thus can possibly be used to treat cancer and genetic
disorders. However, most of the research so far reported has only focused
on their ability as a carrier; almost no research has been reported about the
inherent biological activity of CNTs. On the basis of their previous finding
that SWNTs could selectively stabilize human telomeric i-motif DNA, Chen
et al. found that SWNTs can inhibit telomerase activity and interfere with
telomere functions in cancer cells through stabilization of the i-motif structure.[194] The persistence of i-motifs and the concomitant G-quadruplexes
eventually leads to telomere uncapping and displaces telomere-binding
proteins from telomere. The dysfunctional telomere triggers a DNA damage
response and elicits up-regulation of p16 and p21 proteins. These results
provided new insights into the biomedical effects of SWNTs and highlighted
the potential of SWNTs in cancer therapy.

   In contrast, GO was found to intercalate into double helical DNA due to
their planar, single atomic layer structural motif.[163] The intercalation was
further employed with copper ions to cleave DNA molecules. The DNA cleaving activity of the GO/Cu^{2+} system was attributed to the combination of the
unique structural and chemical features of the GO and high affinity of Cu^{2+}
for DNA. The findings opened a new route to obtaining DNA cleaving systems
and may be useful in improving the specificity of cleavage with certain chemical reagents. DNA cleaving agents are promising candidates for therapeutic
drugs, thus GO combined with metal ions or metal complexes are potential anticancer drugs.[163] As a matter of fact, the combination of GQDs and
anticancer drugs exhibited much higher cytotoxicity and high DNA cleavage
activity *in vitro* (Figure 3.24).[195] Interestedly, the GQDs in the GQDs/antican-
cer drugs system actually played dual roles: they act as anticancer drug carriers and drug enhancers at the same time. The findings revealed the potential
of GQDs in cancer therapy.

3.5 Perspectives

Both CNT and GO have been shown to be versatile platforms for nucleic acid
delivery *in vitro* because of their unique structures, high surface areas, facile functionalized surfaces, and biocompatibility. The encouraging findings
concerning the use of CNT and GO in gene delivery in the last ten years have
been described in the last section.

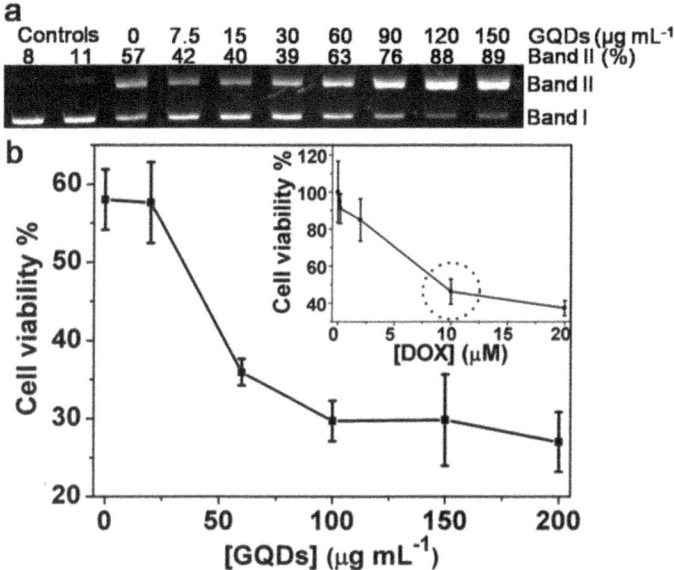

Figure 3.24 GQDs enhanced the activity (a) and the cytotocixity (b) of DOX *in vitro*.[195] Reprinted with permission from ref. 195. Copyright 2013 Nature Publishing Group.

For CNTs, it is crucial to functionalize their surfaces first in order to convert bare CNT, which is insoluble in most solvents, to water-soluble and biocompatible CNT with an improved toxicity profile for its applications. However, different functionalization methods probably produce CNT with different characteristics, which will lead to differences in the properties, and eventually result in different cellular uptake mechanisms, degradation or dissolution, and bioaccumulation. Therefore it's hard to compare their transfection efficiency. Although the CNTs exhibit low toxicity compared to most non-viral gene delivery systems, the toxicological and pharmacological profiles of CNT systems are far less explored. Although many gene delivery studies have been carried out *in vitro*, there are hardly any studies in the literature describing *in vivo* gene expression by CNT-mediated delivery of pDNA. Zhang *et al.* prepared FAM-labeled dsDNA–CNT conjugates, and these conjugates were intratumorally administered in mice bearing Lewis lung carcinoma tumors only for *in vivo* imaging and tracking of the CNT–DNA.[196] Because CNTs are not like small molecules, the administration or implantation of CNTs and their complexes with genes also needs to be explored. Moreover, information regarding how CNTs interact with the cells at the molecular level is lacking.

For GO and its derivatives, because of their inherent surface oxygen-containing functional groups, they exhibit much better biocompatibility than CNTs. However, most GO used in the biomedical research is relatively large, in the region of a couple of hundred nanometers, or even micrometer size. In addition, the large size distribution that originates from the preparation methods could also lead to different effects, especially in gene delivery or drug delivery, since the size may alter their pathway for entering the cells or change their distribution inside the cells. Although the GQDs improved the biocompatibility of GO to a certain extent, because they are much smaller, the information of GQDs interaction with cells at molecule level are still lacking. Continued research into the mechanism of interaction between GO and GQDs with cells at the molecular level may allow us to better envisage their potential and adverse effects as gene or drug carriers, or as anticancer drug enhancers. With this knowledge, it should be possible to rationally design new systems that deliver different types of drugs or gene to the corresponding type of cells, to overcome the resistance problems encountered with anticancer drugs.

References

1. M. M. W. Mooren, D. E. Pulleyblank, S. S. Wijmenga, F. J. M. van de Ven and C. W. Hilbers, *Biochemistry*, 1994, **33**, 7315–7325.
2. P. Belmont, J.-F. Constant and M. Demeunynck, *Chem. Soc. Rev.*, 2001, **30**, 70–81.
3. F. A. Gollmick, M. Lorenz, U. Dornberger, J. von Langen, S. Diekmann and H. Fritzsche, *Nucleic Acids Res.*, 2002, **30**, 2669–2677.
4. H. Kurahashi, H. Inagaki, K. Yamada, T. Ohye, M. Taniguchi, B. S. Emanuel and T. Toda, *J. Biol. Chem.*, 2004, **279**, 35377–35383.
5. F. R. Keene, J. A. Smith and J. G. Collins, *Coord. Chem. Rev.*, 2009, **253**, 2021–2035.
6. J. Choi and T. Majima, *Chem. Soc. Rev.*, 2011, **40**, 5893–5909.
7. T. R. Einert, H. Orland and R. R. Netzl, *Eur. Phys. J. E: Soft Matter Biol. Phys.*, 2011, **34**, 15.
8. I. Dasgupta, X. Gao and G. E. Fox, *Biopolymers*, 2012, **97**, 155–164.
9. M. W. Friederich, E. Vacano and P. J. Hagerman, *Proc. Natl. Acad. Sci. U. S. A.*, 1998, **95**, 3572–3577.
10. J. V. Ditlevson, S. Tornaletti, B. P. Belotserkovskii, V. Teijeiro, G. Wang, K. M. Vasquez and P. C. Hanawalt, *Nucleic Acids Res.*, 2008, **36**, 3163–3170.
11. K. A. M. Ampt, R. M. van der Werf, F. H. T. Nelissen, M. Tessari and S. S. Wijmenga, *Biochemistry*, 2009, **48**, 10499–10508.
12. H. O. Letsch, P. Kück, R. R. Stocsits and B. Misof, *Mol. Biol. Evol.*, 2010, **27**, 2507–2521.
13. B. Duvvuri, V. Duvvuri, J. Wu and G. Wu, *Immunogenetics*, 2012, **64**, 481–496.
14. Y. Sun, E. Atas, L. Lindqvist, N. Sonenberg, J. Pelletier and A. Meller, *Nucleic Acids Res.*, 2012, **40**, 6199–6207.
15. T. A. Brooks, S. Kendrick and L. Hurley, *FEBS J.*, 2010, **277**, 3459–3469.
16. M. Fry and L. A. Loeb, *Proc. Natl. Acad. Sci. U. S. A.*, 1994, **91**, 4950–4954.
17. L. Martorell, K. Johnson, C. A. Boucher and M. Baiget, *Hum. Mol. Genet.*, 1997, **6**, 877–880.
18. M. Gomes-Pereira and D. G. Monckton, *Mutat. Res., Fundam. Mol. Mech. Mutagen.*, 2006, **598**, 15–34.

19. S. M. Mirkin, *Curr. Opin. Struct. Biol.*, 2006, **16**, 351–358.
20. M. Rajeswari, *J. Biosci.*, 2012, **37**, 519–532.
21. A. Rich and S. Zhang, *Nat. Rev. Genet.*, 2003, **4**, 566.
22. B. K. Ray, S. Dhar, A. Shakya and A. Ray, *Proc. Natl. Acad. Sci. U. S. A.*, 2011, **108**, 103–108.
23. A. Rich, D. R. Davies, F. H. C. Crick and J. D. Watson, *J. Mol. Biol.*, 1961, **3**, 71.
24. S. Chakraborty, S. Sharma, P. K. Maiti and Y. Krishnan, *Nucleic Acids Res.*, 2009, **37**, 2810–2817.
25. J. L. Asensio, T. Brown and A. N. Lane, *Structure*, 1999, **7**, 1–11.
26. M. Gellert, M. N. Lipsett and D. R. Davies, *Proc. Natl. Acad. Sci. U. S. A.*, 1962, **48**, 2013–2018.
27. F. W. Kotch, J. C. Fettinger and J. T. Davis, *Org. Lett.*, 2000, **2**, 3277–3280.
28. A. T. Phan, V. Kuryavyi, K. N. Luu and D. J. Patel, *Nucleic Acids Res.*, 2007, **35**, 6517–6525.
29. J. Dai, M. Carver, C. Punchihewa, R. A. Jones and D. Yang, *Nucleic Acids Res.*, 2007, **35**, 4927–4940.
30. T. Li, E. Wang and S. Dong, *J. Am. Chem. Soc.*, 2009, **131**, 15082–15083.
31. J. T. Davis, *Angew. Chem., Int. Ed.*, 2004, **43**, 668–698.
32. S. Zhang, Y. Wu and W. Zhang, *ChemMedChem*, 2014, **9**, 899–911.
33. M. Guéron and J.-L. Leroy, *Curr. Opin. Struct. Biol.*, 2000, **10**, 326–331.
34. K. S. Jin, S. R. Shin, B. Ahn, Y. Rho, S. J. Kim and M. Ree, *J. Phys. Chem. B*, 2009, **113**, 1852–1856.
35. J.-L. Mergny, L. Lacroix, X. Han, J.-L. Leroy and C. Helene, *J. Am. Chem. Soc.*, 1995, **117**, 8887–8898.
36. S. Kendrick, Y. Akiyama, S. M. Hecht and L. H. Hurley, *J. Am. Chem. Soc.*, 2009, **131**, 17667–17676.
37. D. Miyoshi, H. Karimata and N. Sugimoto, *J. Am. Chem. Soc.*, 2006, **128**, 7957–7963.
38. J. Zhou, C. Wei, G. Jia, X. Wang, Z. Feng and C. Li, *Mol. BioSyst.*, 2010, **6**, 580–586.
39. O. Y. Fedoroff, A. Rangan, V. V. Chemeris and L. H. Hurley, *Biochemistry*, 2000, **39**, 15083–15090.
40. X. Li, Y. Peng, J. Ren and X. Qu, *Proc. Natl. Acad. Sci. U. S. A.*, 2006, **103**, 19658–19663.
41. S. R. Shin, K. S. Jin, C. K. Lee, S. I. Kim, G. M. Spinks, I. So, J.-H. Jeon, T. M. Kang, J. Y. Mun, S.-S. Han, M. Ree and S. J. Kim, *Adv. Mater.*, 2009, **21**, 1907–1910.
42. A. T. Phan, M. Guéron and J.-L. Leroy, *J. Mol. Biol.*, 2000, **299**, 123–144.
43. *Small Molecule DNA and RNA Binders: From Synthesis to Nucleic Acid Complexes*. Wiley-VCH Verlag GmbH & Co. KGaA 2004.
44. G. Tsuji, K. Kawakami and S. Sasaki, *Bioorg. Med. Chem.*, 2013, **21**, 6063–6068.
45. B. Spingler and C. Da Pieve, *Dalton Trans.*, 2005, 1637–1643.
46. Z. Wu, T. Tian, J. Yu, X. Weng, Y. Liu and X. Zhou, *Angew. Chem., Int. Ed.*, 2011, **50**, 11962–11967.

47. J. K. Choi, A. D'Urso and M. Balaz, *J. Inorg. Biochem.*, 2013, **127**, 1–6.
48. N. Morii, G. Kido, T. Konakahara and H. Morii, *Biomacromolecules*, 2005, **6**, 3259–3266.
49. A. Medina-Molner and B. Spingler, *Chem. Commun.*, 2012, **48**, 1961–1963.
50. T. F. Kagawa, B. H. Geierstanger, A. H. Wang and P. S. Ho, *J. Biol. Chem.*, 1991, **266**, 20175–20184.
51. Y.-G. Gao, M. Sriram and A. H. J. Wang, *Nucleic Acids Res.*, 1993, **21**, 4093–4101.
52. B. Spingler and P. M. Antoni, *Chem.–Eur. J.*, 2007, **13**, 6617–6622.
53. M. Filimonova, V. Gubskaya, R. Davidov, A. Garusov and I. d. Nuretdinov, *Int. J. Biol. Macromol.*, 2008, **43**, 289–294.
54. T. Chatake and T. Sunami, *J. Inorg. Biochem.*, 2013, **124**, 15–25.
55. P. Giri and G. S. Kumar, *Mol. BioSyst.*, 2010, **6**, 81–88.
56. K.-Y. Zee-Cheng, K. D. Paull and C. C. Cheng, *J. Med. Chem.*, 1974, **17**, 347–351.
57. B. Gatto, M. M. Sanders, C. Yu, H.-Y. Wu, D. Makhey, E. J. LaVoie and L. F. Liu, *Cancer Res.*, 1996, **56**, 2795–2800.
58. Ö. Persil, C. T. Santai, S. S. Jain and N. V. Hud, *J. Am. Chem. Soc.*, 2004, **126**, 8644–8645.
59. I. S. Joung, Ö. Persil Çetinkol, N. V. Hud and T. E. Cheatham, *Nucleic Acids Res.*, 2009, **37**, 7715–7727.
60. Ö. P. Çetinkol and N. V. Hud, *Nucleic Acids Res.*, 2009, **37**, 611–621.
61. W. D. Wilson, F. A. Tanious, S. Mizan, S. Yao, A. S. Kiselyov, G. Zon and L. Strekowski, *Biochemistry*, 1993, **32**, 10614–10621.
62. K. R. Fox, *Curr. Med. Chem.*, 2000, 7, 17–37.
63. J. L. Mergny, G. Duval-Valentin, C. H. Nguyen, L. Perrouault, B. Faucon, M. Rougée, T. Montenay-Garestier, E. Bisagni and C. Hélène, *Science*, 1992, **256**, 1681–1684.
64. L. J. P. Latimer, N. Payton, G. Forsyth and J. S. Lee, *Biochem. Cell Biol.*, 1995, **73**, 11–18.
65. M. Polak and N. V. Hud, *Nucleic Acids Res.*, 2002, **30**, 983–992.
66. S. A. Cassidy, L. Strekowski and K. R. Fox, *Nucleic Acids Res.*, 1996, **24**, 4133–4138.
67. M. Keppler, O. Zegrocka, L. Strekowski and K. R. Fox, *FEBS Lett.*, 1999, **447**, 223–226.
68. M. D. Keppler, P. L. James, S. Neidle, T. Brown and K. R. Fox, *Eur. J. Biochem.*, 2003, **270**, 4982–4992.
69. Y. Kan, B. Armitage and G. B. Schuster, *Biochemistry*, 1997, **36**, 1461–1466.
70. Y. Kan and G. B. Schuster, *J. Am. Chem. Soc.*, 1999, **121**, 11607–11614.
71. M. Sato, T. Moriguchi and K. Shinozuka, *Bioorg. Med. Chem. Lett.*, 2004, **14**, 1305–1308.
72. A. T. M. Z. Azam, T. Moriguchi and K. Shinozuka, *Chem. Commun.*, 2006, 335–337.
73. T. Moriguchi, A. T. M. Z. Azam and K. Shinozuka, *Bioconjugate Chem.*, 2011, **22**, 1039–1045.
74. S. Tsukahara, S. G. Kim and H. Takaku, *Biochem. Biophys. Res. Commun.*, 1993, **196**, 990–996.

75. J. Selvasekaran and K. D. Turnbull, *Nucleic Acids Res.*, 1999, **27**, 624–627.
76. A. K. Jain and S. Bhattacharya, *Bioconjugate Chem.*, 2010, **21**, 1389–1403.
77. J. Robles and L. W. McLaughlin, *J. Am. Chem. Soc.*, 1997, **119**, 6014–6021.
78. A. K. Jain, S. K. Awasthi and V. Tandon, *Bioorg. Med. Chem.*, 2006, **14**, 6444–6452.
79. D. P. Arya, *Acc. Chem. Res.*, 2010, **44**, 134–146.
80. N. W. Kim, M. A. Piatyszek, K. R. Prowse, C. B. Harley, M. D. West, P. L. Ho, G. M. Coviello, W. E. Wright, S. L. Weinrich and J. W. Shay, *Science*, 1994, **266**, 2011–2015.
81. G. B. Morin, *J. Natl. Cancer Inst.*, 1995, **87**, 859–861.
82. E. K. Parkinson, *Br. J. Cancer*, 1996, **73**, 1–4.
83. E. Izbicka, R. T. Wheelhouse, E. Raymond, K. K. Davidson, R. A. Lawrence, D. Sun, B. E. Windle, L. H. Hurley and D. D. Von Hoff, *Cancer Res.*, 1999, **59**, 639–644.
84. H. Mita, T. Ohyama, Y. Tanaka and Y. Yamamoto, *Biochemistry*, 2006, **45**, 6765–6772.
85. A. T. Phan, V. Kuryavyi, H. Y. Gaw and D. J. Patel, *Nat. Chem. Biol.*, 2005, **1**, 167–173.
86. L. R. Keating and V. A. Szalai, *Biochemistry*, 2004, **43**, 15891–15900.
87. I. M. Dixon, F. Lopez, J.-P. Estève, A. M. Tejera, M. A. Blasco, G. Pratviel and B. Meunier, *ChemBioChem*, 2005, **6**, 123–132.
88. S. Bianco, C. Musetti, A. Waldeck, S. Sparapani, J. D. Seitz, A. P. Krapcho, M. Palumbo and C. Sissi, *Dalton Trans.*, 2010, **39**, 5833–5841.
89. P. Wang, C.-H. Leung, D.-L. Ma, S.-C. Yan and C.-M. Che, *Chem.–Eur. J.*, 2010, **16**, 6900–6911.
90. V. Pradines and G. Pratviel, *Angew. Chem., Int. Ed.*, 2013, **52**, 2185–2188.
91. K. E. Erkkila, D. T. Odom and J. K. Barton, *Chem. Rev.*, 1999, **99**, 2777–2796.
92. T. Biver, F. Secco and M. Venturini, *Coord. Chem. Rev.*, 2008, **252**, 1163–1177.
93. A. Rescifina, C. Zagni, M. G. Varrica, V. Pistarà and A. Corsaro, *Eur. J. Med. Chem.*, 2014, **74**, 95–115.
94. N. V. Anantha, M. Azam and R. D. Sheardy, *Biochemistry*, 1998, **37**, 2709–2714.
95. I. Haq, J. O. Trent, B. Z. Chowdhry and T. C. Jenkins, *J. Am. Chem. Soc.*, 1999, **121**, 1768–1779.
96. A. Randazzo, A. Galeone and L. Mayol, *Chem. Commun.*, 2001, 1030–1031.
97. G. N. Parkinson, F. Cuenca and S. Neidle, *J. Mol. Biol.*, 2008, **381**, 1145–1156.
98. R. Rodriguez, G. D. Pantoş, D. P. N. Gonçalves, J. K. M. Sanders and S. Balasubramanian, *Angew. Chem.*, 2007, **119**, 5501–5503.
99. H. A. Day, P. Pavlou and Z. A. E. Waller, *Bioorg. Med. Chem.*, 2014, **22**, 4407–4418.
100. S. Fernández, R. Eritja, A. Aviñó, J. Jaumot and R. Gargallo, *Int. J. Biol. Macromol.*, 2011, **49**, 729–736.
101. C. L. Grand, H. Han, R. M. Muñoz, S. Weitman, D. D. Von Hoff, L. H. Hurley and D. J. Bearss, *Mol. Cancer Ther.*, 2002, **1**, 565–573.

102. I. J. Lee, J. W. Yi and B. H. Kim, *Chem. Commun.*, 2009, 5383–5385.
103. S. Iijima, *Nature*, 1991, **354**, 56–58.
104. T. Fujigaya and N. Nakashima, in *Chemistry of Nanocarbons*, John Wiley & Sons, Ltd, 2010, pp. 301–331.
105. F. Bonaccorso, P.-H. Tan and A. C. Ferrari, *ACS Nano*, 2013, **7**, 1838–1844.
106. A. Javey, J. Guo, Q. Wang, M. Lundstrom and H. Dai, *Nature*, 2003, **424**, 654–657.
107. Q. Cao and J. Rogers, *Nano Res.*, 2008, **1**, 259–272.
108. P. A. Gordon and R. B. Saeger, *Ind. Eng. Chem. Res.*, 1999, **38**, 4647–4655.
109. H. Ago, K. Petritsch, M. S. P. Shaffer, A. H. Windle and R. H. Friend, *Adv. Mater.*, 1999, **11**, 1281–1285.
110. M. Baxendale, *IEE Proc.: Nanobiotechnol.*, 2003, **150**, 3.
111. C. R. Martin and P. Kohli, *Nat. Rev. Drug Discovery*, 2003, **2**, 29.
112. Z. Liu, S. Tabakman, K. Welsher and H. Dai, *Nano Res.*, 2009, **2**, 85–120.
113. M. Bottini, N. Rosato and N. Bottini, *Biomacromolecules*, 2011, **12**, 3381–3393.
114. X. Wang and Z. Liu, *Chin. Sci. Bull.*, 2012, **57**, 167–180.
115. M. Zheng, A. Jagota, E. D. Semke, B. A. Diner, R. S. McLean, S. R. Lustig, R. E. Richardson and N. G. Tassi, *Nat. Mater.*, 2003, **2**, 338–342.
116. M. E. Hughes, E. Brandin and J. A. Golovchenko, *Nano Lett.*, 2007, **7**, 1191–1194.
117. O. P. R. G. I. Dovbeshko, E. D. Obraztsova, Y. V. Shtogun and E. O. Andreev, *Study of DNA Interaction with Carbon Nanotubes*, 2003.
118. J. F. Campbell, I. Tessmer, H. H. Thorp and D. A. Erie, *J. Am. Chem. Soc.*, 2008, **130**, 10648–10655.
119. S. Meng, P. Maragakis, C. Papaloukas and E. Kaxiras, *Nano Lett.*, 2006, **7**, 45–50.
120. D. A. Yarotski, S. V. Kilina, A. A. Talin, S. Tretiak, O. V. Prezhdo, A. V. Balatsky and A. J. Taylor, *Nano Lett.*, 2009, **9**, 12–17.
121. F. F. Contreras-Torres and E. Martínez-Lorán, *Wiley Interdiscip. Rev.: Comput. Mol. Sci.*, 2011, **1**, 902–919.
122. Y. Noguchi, T. Fujigaya, Y. Niidome and N. Nakashima, *Chem. Phys. Lett.*, 2008, **455**, 249–251.
123. H. Gao, Y. Kong, D. Cui and C. S. Ozkan, *Nano Lett.*, 2003, **3**, 471–473.
124. G. Huajian and K. Yong, *Annu. Rev. Mater. Res.*, 2004, **34**, 123.
125. A. Majumder, M. Khazaee, J. Opitz, E. Beyer, L. Baraban and G. Cuniberti, *Phys. Chem. Chem. Phys.*, 2013, **15**, 17158–17164.
126. M. Cha, S. Jung, M.-H. Cha, G. Kim, J. Ihm and J. Lee, *Nano Lett.*, 2009, **9**, 1345–1349.
127. S. S. Kim, C. L. Hisey, Z. Kuang, D. A. Comfort, B. L. Farmer and R. R. Naik, *Nanoscale*, 2013, **5**, 4931–4936.
128. B. Maji, S. K. Samanta and S. Bhattacharya, *Nanoscale*, 2014, **6**, 3721–3730.
129. A. Fernando, E. H. Mary, A. G. Jene and B. Daniel, *Nanotechnology*, 2009, **20**, 395101.

130. Z. Xiao, X. Wang, X. Xu, H. Zhang, Y. Li and Y. Wang, *J. Phys. Chem. C*, 2011, **115**, 21546–21558.
131. D. Roxbury, A. Jagota and J. Mittal, *J. Am. Chem. Soc.*, 2011, **133**, 13545–13550.
132. D. A. Heller, E. S. Jeng, T.-K. Yeung, B. M. Martinez, A. E. Moll, J. B. Gastala and M. S. Strano, *Science*, 2006, **311**, 508–511.
133. X. Li, Y. Peng and X. Qu, *Nucleic Acids Res.*, 2006, **34**, 3670–3676.
134. D. R. Dreyer, S. Park, C. W. Bielawski and R. S. Ruoff, *Chem. Soc. Rev.*, 2010, **39**, 228–240.
135. D. Chen, H. Feng and J. Li, *Chem. Rev.*, 2012, **112**, 6027–6053.
136. J. Liu, L. Cui and D. Losic, *Acta Biomater.*, 2013, **9**, 9243–9257.
137. C. Chung, Y.-K. Kim, D. Shin, S.-R. Ryoo, B. H. Hong and D.-H. Min, *Acc. Chem. Res.*, 2013, **46**, 2211–2224.
138. J.-L. Li, B. Tang, B. Yuan, L. Sun and X.-G. Wang, *Biomaterials*, 2013, **34**, 9519–9534.
139. L. Gao, C. Lian, Y. Zhou, L. Yan, Q. Li, C. Zhang, L. Chen and K. Chen, *Biosens. Bioelectron.*, 2014, **60**, 22–29.
140. Y. Zhang, C. Wu, S. Guo and J. Zhang, *Nanotechnol. Rev.*, 2013, **2**, 27.
141. C. Xu, X. Wang and J. Zhu, *J. Phys. Chem. C*, 2008, **112**, 19841–19845.
142. J. R. Lomeda, C. D. Doyle, D. V. Kosynkin, W.-F. Hwang and J. M. Tour, *J. Am. Chem. Soc.*, 2008, **130**, 16201–16206.
143. K. Jasuja and V. Berry, *ACS Nano*, 2009, **3**, 2358–2366.
144. J. Zhang, F. Zhang, H. Yang, X. Huang, H. Liu, J. Zhang and S. Guo, *Langmuir*, 2010, **26**, 6083–6085.
145. J. Zhang, G. Shen, W. Wang, X. Zhou and S. Guo, *J. Mater. Chem.*, 2010, **20**, 10824–10828.
146. F. Zhang, B. Zheng, J. Zhang, X. Huang, H. Liu, S. Guo and J. Zhang, *J. Phys. Chem. C*, 2010, **114**, 8469–8473.
147. D. Cheng, G. Hong, W. Wang, R. Yuan, H. Ai, J. Shen, B. Liang, J. Gao and X. Shuai, *J. Mater. Chem.*, 2011, **21**, 4796–4804.
148. H. Wu, G. Gao, X. Zhou, Y. Zhang and S. Guo, *CrystEngComm*, 2012, **14**, 499–504.
149. S. He, B. Song, D. Li, C. Zhu, W. Qi, Y. Wen, L. Wang, S. Song, H. Fang and C. Fan, *Adv. Funct. Mater.*, 2010, **20**, 453–459.
150. Z. Tang, H. Wu, J. R. Cort, G. W. Buchko, Y. Zhang, Y. Shao, I. A. Aksay, J. Liu and Y. Lin, *Small*, 2010, **6**, 1205–1209.
151. P.-J. Huang and J. Liu, *Nanomaterials*, 2013, **3**, 221–228.
152. Y. Zhang, Y. Liu, S. J. Zhen and C. Z. Huang, *Chem. Commun.*, 2011, **47**, 11718–11720.
153. Q. Zhu, D. Xiang, C. Zhang, X. Ji and Z. He, *Analyst*, 2013, **138**, 5194–5196.
154. P.-J. Huang and J. Liu, *Nanomaterials*, 2013, **3**, 221–228.
155. Y. He, G. Huang and H. Cui, *ACS Appl. Mater. Interfaces*, 2013, **5**, 11336–11340.
156. B. Liu, Z. Sun, X. Zhang and J. Liu, *Anal. Chem.*, 2013, **85**, 7987–7993.
157. H. Zhang, S. Jia, M. Lv, J. Shi, X. Zuo, S. Su, L. Wang, W. Huang, C. Fan and Q. Huang, *Anal. Chem.*, 2014, **86**, 4047–4051.

158. E. Yoo, J. Kim, E. Hosono, H.-S. Zhou, T. Kudo and I. Honma, *Nano Lett.*, 2008, **8**, 2277–2282.
159. B. G. Choi, H. Park, T. J. Park, M. H. Yang, J. S. Kim, S.-Y. Jang, N. S. Heo, S. Y. Lee, J. Kong and W. H. Hong, *ACS Nano*, 2010, **4**, 2910–2918.
160. H. Lei, L. Mi, X. Zhou, J. Chen, J. Hu, S. Guo and Y. Zhang, *Nanoscale*, 2011, **3**, 3888–3892.
161. M. Liu, H. Zhao, S. Chen, H. Yu and X. Quan, *Chem. Commun.*, 2012, **48**, 564–566.
162. J. S. Park, N.-I. Goo and D.-E. Kim, *Langmuir*, 2014, **30**, 12587–12595.
163. H. Ren, C. Wang, J. Zhang, X. Zhou, D. Xu, J. Zheng, S. Guo and J. Zhang, *ACS Nano*, 2010, **4**, 7169–7174.
164. J. Shen, Y. Zhu, X. Yang and C. Li, *Chem. Commun.*, 2012, **48**, 3686–3699.
165. M. Bacon, S. J. Bradley and T. Nann, *Part. Part. Syst. Charact.*, 2014, **31**, 415–428.
166. X. Zhou, Y. Zhang, C. Wang, X. Wu, Y. Yang, B. Zheng, H. Wu, S. Guo and J. Zhang, *ACS Nano*, 2012, **6**, 6592–6599.
167. X. Chen, X. Zhou, T. Han, J. Wu, J. Zhang and S. Guo, *ACS Nano*, 2012, 7, 531–537.
168. Y. Peng, X. Li, J. Ren and X. Qu, *Chem. Commun.*, 2007, 5176–5178.
169. C. Zhao, J. Ren and X. Qu, *Chem.–Eur. J.*, 2008, **14**, 5435–5439.
170. D. Pantarotto, R. Singh, D. McCarthy, M. Erhardt, J.-P. Briand, M. Prato, K. Kostarelos and A. Bianco, *Angew. Chem., Int. Ed.*, 2004, **43**, 5242–5246.
171. R. Singh, D. Pantarotto, D. McCarthy, O. Chaloin, J. Hoebeke, C. D. Partidos, J.-P. Briand, M. Prato, A. Bianco and K. Kostarelos, *J. Am. Chem. Soc.*, 2005, **127**, 4388–4396.
172. Y. Liu, D.-C. Wu, W.-D. Zhang, X. Jiang, C.-B. He, T. S. Chung, S. H. Goh and K. W. Leong, *Angew. Chem., Int. Ed.*, 2005, **44**, 4782–4785.
173. B. Behnam, W. T. Shier, A. H. Nia, K. Abnous and M. Ramezani, *Int. J. Pharm.*, 2013, **454**, 204–215.
174. L. Gao, L. Nie, T. Wang, Y. Qin, Z. Guo, D. Yang and X. Yan, *ChemBioChem*, 2006, **7**, 239–242.
175. M. Liu, B. Chen, Y. Xue, J. Huang, L. Zhang, S. Huang, Q. Li and Z. Zhang, *Bioconjugate Chem.*, 2011, **22**, 2237–2243.
176. W. Qin, K. Yang, H. Tang, L. Tan, Q. Xie, M. Ma, Y. Zhang and S. Yao, *Colloids Surf., B*, 2011, **84**, 206–213.
177. M. Ahmed, X. Jiang, Z. Deng and R. Narain, *Bioconjugate Chem.*, 2009, **20**, 2017–2022.
178. A. Karmakar, S. M. Bratton, E. Dervishi, A. Ghosh, M. Mahmood, Y. Xu, L. Saeed, T. Mustafa, D. Casciano, A. Radominska-Pandya and A. S. Biris, *Int. J. Nanomed.*, 2011, **6**, 1045–1055.
179. V. Sanz, C. Tilmacîu, B. Soula, E. Flahaut, H. M. Coley, S. R. P. Silva and J. McFadden, *Carbon*, 2011, **49**, 5348–5358.
180. F. M. P. Tonelli, S. M. S. N. Lacerda, M. A. Silva, E. S. Avila, L. O. Ladeira, L. R. Franca and R. R. Resende, *RSC Adv.*, 2014, **4**, 37985–37987.
181. C. Geyik, S. Evran, S. Timur and A. Telefoncu, *Biotechnol. Prog.*, 2014, **30**, 224–232.

182. A. Cifuentes-Rius, A. de Pablo, V. Ramos-Pérez and S. Borrós, *Plasma Processes Polym.*, 2014, **11**, 704–713.
183. D.-H. Jung, B. H. Kim, Y. T. Lim, J. Kim, S. Y. Lee and H.-T. Jung, *Carbon*, 2010, **48**, 1070–1078.
184. L. Feng, S. Zhang and Z. Liu, *Nanoscale*, 2011, **3**, 1252–1257.
185. B. Chen, M. Liu, L. Zhang, J. Huang, J. Yao and Z. Zhang, *J. Mater. Chem.*, 2011, **21**, 7736–7741.
186. H. Kim, R. Namgung, K. Singha, I.-K. Oh and W. J. Kim, *Bioconjugate Chem.*, 2011, **22**, 2558–2567.
187. T. Ren, L. Li, X. Cai, H. Dong, S. Liu and Y. Li, *Polym. Chem.*, 2012, **3**, 2561–2569.
188. J. Zhang, L. Feng, X. Tan, X. Shi, L. Xu, Z. Liu and R. Peng, *Part. Part. Syst. Charact.*, 2013, **30**, 794–803.
189. K. Li, L. Feng, J. Shen, Q. Zhang, Z. Liu, S.-T. Lee and J. Liu, *ACS Appl. Mater. Interfaces*, 2014, **6**, 5900–5907.
190. X. Yang, N. Zhao and F.-J. Xu, *Nanoscale*, 2014, **6**, 6141–6150.
191. H. Bao, Y. Pan, Y. Ping, N. G. Sahoo, T. Wu, L. Li, J. Li and L. H. Gan, *Small*, 2011, **7**, 1569–1578.
192. H. Hu, C. Tang and C. Yin, *Mater. Lett.*, 2014, **125**, 82–85.
193. X. Yang, G. Niu, X. Cao, Y. Wen, R. Xiang, H. Duan and Y. Chen, *J. Mater. Chem.*, 2012, **22**, 6649–6654.
194. Y. Chen, K. Qu, C. Zhao, L. Wu, J. Ren, J. Wang and X. Qu, *Nat. Commun.*, 2012, **3**, 1074.
195. C. Wang, C. Wu, X. Zhou, T. Han, X. Xin, J. Wu, J. Zhang and S. Guo, *Sci. Rep.*, 2013, **3**, 2852–2860.
196. Z. Zhang, X. Yang, Y. Zhang, B. Zeng, S. Wang, T. Zhu, R. B. S. Roden, Y. Chen and R. Yang, *Clin. Cancer Res.*, 2006, **12**, 4933–4939.
197. L. Martino, A. Virno, B. Pagano, A. Virgilio, S. Di Micco, A. Galeone, C. Giancola, G. Bifulco, L. Mayol and A. Randazzo, *J. Am. Chem. Soc.*, 2007, **129**, 16048–16056.

CHAPTER 4

Chemical Biotechnology of In Vitro Synthetic Biosystems for Biomanufacturing

ZHIGUANG ZHU[a] AND YI-HENG PERCIVAL ZHANG*[a,b,c]

[a]Cell-Free Bioinnovations Inc., 2200 Kraft Drive, Suite 1200B, Blacksburg, Virginia 24060, USA; [b]Biological Systems Engineering Department, Virginia Tech, 304 Seitz Hall, Blacksburg, Virginia 24061, USA; [c]Institute for Critical Technology and Applied Science (ICTAS), Virginia Tech, Blacksburg, Virginia 24061, USA
*E-mail: ypzhang@vt.edu

4.1 Introduction

4.1.1 Definition

Biocatalysis and biotransformation refers to the use of biocatalysts, such as enzymes or whole cells, to perform the synthesis, conversion, or degradation of chemical species.[1] It has been increasingly used in the pharmaceutical, chemical, food, energy, agricultural, and environmental industries because of its high catalytic activity under mild conditions, exquisite chemo-, regio-, and enantio-selectivities, and great sustainability without costly noble or rare metals.[2] Biocatalysts can be classified into whole cells, cell extracts, single enzymes, or enzyme cocktails (Figure 4.1).[3,4] Whole cells, especially microbes, have been utilized in the production of fermented food, beer,

RSC Green Chemistry No. 34
Chemical Biotechnology and Bioengineering
By Xuhong Qian, Zhenjiang Zhao, Yufang Xu, Jianhe Xu, Y.-H. Percival Zhang, Jingyan Zhang, Yangchun Yong, and Fengxian Hu
© X.-H. Qian, Z.-J. Zhao, Y.-F. Xu, J.-H. Xu, Y.-H. P. Zhang, J.-Y. Zhang, Y.-C. Yong and F.-X. Hu, 2015
Published by the Royal Society of Chemistry, www.rsc.org

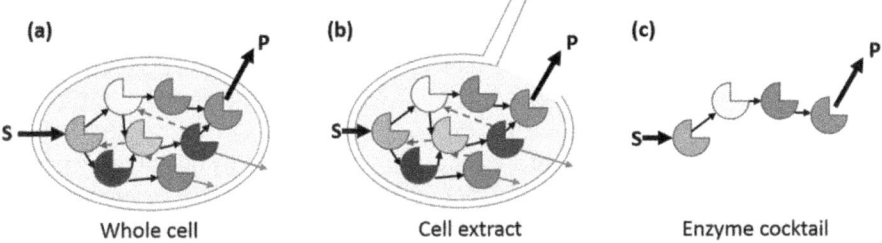

Figure 4.1 Three representative biocatalysts: (a) whole cell, (b) cell extract, and (c) enzyme cocktail.

wines, drugs, chemicals, and so forth to meet mankind's myriad needs for thousands of years.[5-7] Single enzymes have been commercialized for more than a half century for the production of fructose, chemicals, and semi-synthetic antibiotics. The use of cell extracts for the production of high-value vaccines, vitamins, and proteins has been studied in the last two decades.[8-10] Compared to whole cell systems, cell extracts eliminate the barriers of cell walls and cell membranes, creating more accessible and controllable, but probably vulnerable, environments. These three systems are well-known and their fundamentals and applications have been reviewed elsewhere.[2,11-15]

In vitro biosystems, also called cell-free synthetic pathway biotransformations, enzyme cocktails, synthetic cascade enzyme factories, synthetic cascade manufacturing, synthetic biochemistry, and so on, are the *in vitro* assembly of a number of enzymes, which may be isolated from different organisms, and/or coenzymes, for the production of desired products that may not be produced by microbes or abiotic catalysts.[4] Because the primary goal of such *in vitro* biosystems is the manufacturing of the desired products rather than the investigation of natural biochemical pathways, these biosystems are regarded as a new biomanufacturing platform. This concept is different from the traditional enzymatic conversion that is mediated by one or several enzymes only, but it can be regarded as an extension of traditional enzyme biocatalysis. Moreover, *in vitro* biosystems, like multi-enzyme systems, have more advantages over single-enzyme biocatalysts, such as fewer operation steps, smaller reaction volumes, higher volumetric and space-time yields, shorter cycle times, less waste generation, removal of product inhibition, driving reaction equilibria, and the capability of performing unnatural pathways.[16-18] Because numerous excellent reviews on enzyme-based biocatalysis have been published elsewhere,[1,19] here we focus on *in vitro* biosystems for biomanufacturing with special emphasis on meeting bioprocessing needs.

4.1.2 History

In vitro biosystems have been used to investigate fundamental aspects of natural biochemical pathways for more than a century. As early as 1897, Eduard Buchner discovered the conversion of glucose to ethanol using a cell-free yeast extract and he was awarded the Nobel Chemistry Prize ten years later.[20] In the 1960s, several industrial enzymes were used to produce desired products rather than whole cells, such as fructose from glucose. Over time, enzyme-based biocatalysis has evolved to become more complicated and has more applications. For example, the synthesis of chiral alcohol can be conducted *via in situ* cofactor regeneration mediated by coupled enzymes, such as hydrogenase and alcohol dehydrogenase, or mono-oxygenase and formate dehydrogenase.[21] A few enzymes, including endoglucanase, cellobiohydrolase, and β-glucosidase, work synergistically to degrade recalcitrant cellulose into glucose.[22] Moreover, the biosynthesis of hydroxylated products has been demonstrated by using aldolase, transketolase, oxynitrilase, and other related enzymes.[23,24] More advances have been made using such one-pot multi-enzyme strategies to synthesize glycopeptides,[25] uridine diphosphate sugars,[26] and aminopolyols.[27]

The use of isolated enzymes to reconstitute natural or non-natural pathways for the purpose of industrial production emerged in the 2000s.[28–30] Unlike multi-enzymes in one pot, which normally involves two or three enzymes performing certain relatively simple bioconversions, a synthetic enzymatic pathway usually involves more than ten enzymes to mimic natural pathways or form non-natural pathways for implementing very complicated bioconversions.[31] As the system's complexity increases with more enzymes involved, more considerations and tradeoffs about experimental conditions, such as pH, temperature, metal ions, and buffer, should be taken into account. Now, *in vitro* biosystems are a promising biomanufacturing platform for the production of numerous high-value compounds.[32,33] For example, fast synthesis of *in vivo* cytotoxic, regulatory, or unstable proteins that are difficult to be expressed in living cells has been demonstrated using cell-free protein synthesis systems.[13,34] In the pharmaceutical and fine chemical industries, more manufacturers are starting to use cell-free systems to produce desired stereoisomers and enantiomerically pure amino acids, amines, amides, nitriles, alcohols, and organic acids.[2] In these cases, due to the high prices of these products varying from millions to billions of dollars per kilogram, their production has less stringent requirements on the cost of enzymes and cofactors, their stability, or product yield, but has a high requirement on product titer.

In vitro biosystems comprised of synthetic enzymatic pathways are proposed to be a future low-cost biomanufacturing platform suitable for the production of low-selling price biocommodities, including biofuels, biochemicals, bioplastics, food, and feed.[30] The selling prices of biocommodities range from less than one dollar to several dollars per kilogram, where their prices are dominated by the substrate costs. Therefore, product yield, reaction rate, biocatalyst cost, and production separation cost are critically important factors for the economic viability of biomanufacturing.[17] Proof-of-concept experiments based on *in vitro* synthetic enzymatic pathways have been demonstrated to produce hydrogen,[35–37] liquid biofuels (*e.g.*, ethanol, isobutanol, *n*-butanol),[38,39] bioelectricity,[40–42] polyols (*e.g.*, xylitol),[43] and carbohydrates (*e.g.*, starch and fructose),[44,45] suggesting great industrial potential in the near future.

4.1.3 Advantages and Disadvantages

In vitro biosystems for biomanufacturing feature several industrial production advantages over whole-cell-based biomanufacturing.[30,46-48] Firstly, high product yields are accomplished by the elimination of both side reactions and the synthesis of cell mass. Living microbes have to spend extra energy and resources on self-duplication, the maintenance of basic metabolism, and the production of undesired products, leading to low product yields, especially for aerobic fermentations. Secondly, fast reaction rates can be achieved due to the better mass transfer without the barrier of cell membranes. Another reason for the fast reaction rates is that enzymes have much larger turnover numbers than the same volume or the weight of microorganisms.[49] Thirdly, enzymes usually tolerate toxins and solvents much better than whole cells because of a lack of labile cell membrane. Fourthly, the reconstitution of synthetic enzymatic pathways can implement some non-natural reactions that can never occur in living cells. Fifthly, the reaction equilibrium may be shifted in favor of product formation through well-designed synthetic enzymatic pathways. Last but not the least, it is easier to manipulate and optimize the ratios, amounts, spatial locations, and activities of isolated enzymes and pathways including cofactors than those in the living organisms, which involve complicated cell metabolism. The standardization of building blocks for cell-free biosystems is much easier than *in vivo* systems.

In vitro biosystems for biomanufacturing are still in their infant stage compared to whole cell fermentation. The greatest doubt regarding *in vitro* biosystems comes from the high production cost of many enzymes and poor enzyme stabilities.[50] This doubt is normally based on the common paradigm that high-purity enzymes used in academic labs and the pharmaceutical industry are expensive and unstable. Indeed, the production costs of industrial bulk enzymes have decreased to as low as several US dollars per kilogram of dry protein. On the other hand, enzyme stability can be improved greatly by the discovery of more thermostable enzymes, engineering enzymes with increased activities and stabilities by protein engineering, and the immobilization of enzymes.[51-53] In addition, enzyme engineering could solve the issue of the tradeoff between the optimal conditions of numerous enzymes. The costs and stabilities of natural cofactors (*e.g.*, ATP and NAD(P)) may be another critical problem for the development of *in vitro* biosystems.[21] Besides the careful design of cofactor-free or balanced cofactor pathways, the use of low-cost and stable biomimetic cofactors that have similar functions is a promising alternative. By considering its unique bioprocessing advantages and special application needs, *in vitro* synthetic cascade enzyme biomanufacturing will become a new disruptive manufacturing platform in the near future.

4.2 Biomanufacturing Advantages of *In Vitro* Synthetic Biosystems

4.2.1 High Product Yield

High product yield is one of the most striking features of cell-free biosystems. For instance, the highest yield of hydrogen can be produced from a number of carbohydrate substrates *via* synthetic enzymatic pathways (Figure 4.2).[35-37,54,55] Hydrogen is one of the most important commodity chemicals with a global market of approximately 100 billion US dollars and will be a future transportation energy carrier with a projected market size of approximately 10 trillion US dollars. However, whole-cell microorganisms generate a maximum of 4 molecules of dihydrogen per glucose unit, which is called the Thauer limit.[56] To break this limit, synthetic enzymatic pathways containing more than 10 enzymes have been designed, with a theoretical yield of nearly 12 molecules of hydrogen per glucose unit (or 10 hydrogen molecules per xylose unit) achieved. To convert these sugars into hydrogen, synthetic enzymatic pathways can be divided into four steps: (1) the generation of phosphorylated monomer sugars (*e.g.*, glucose 6-phosphate (G6P)); (2) the generation of 2 NADPH from G6P *via* an oxidative pentose phosphate pathway; (3) hydrogen production catalyzed by the hydrogenase, where the reducing power comes from the regeneration of NADPH by two dehydrogenases in an oxidative pentose phosphate pathway; and (4) the recycling of ribulose 5-phosphate to G6P through a non-oxidative pentose phosphate pathway.

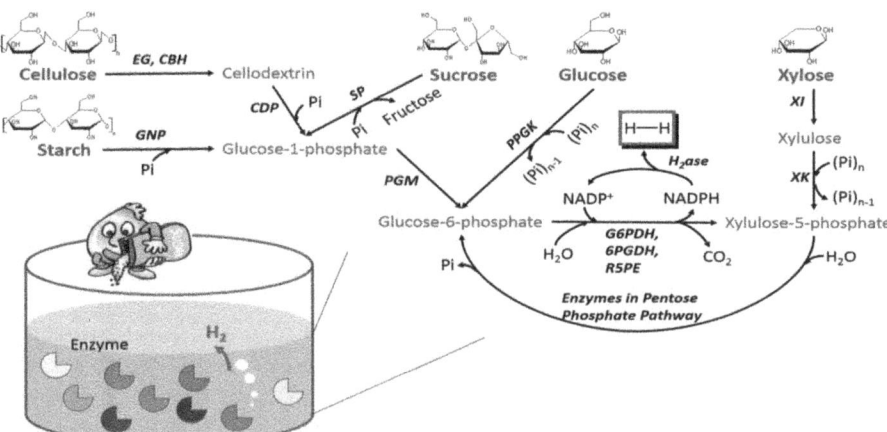

Figure 4.2   Enzymatic hydrogen production from various carbohydrates *via in vitro* synthetic enzymatic pathways. EG: endoglucanase, CBH: cellobiohydrolase, GNP: glucan phosphorylase, CDP: cellodextrin phosphorylase, SP: sucrose phosphorylase, PGM: phosphoglucomutase, PPGK: polyphosphate glucokinase, G6PDH: glucose-6-phosphate dehydrogenase, 6PGDH: 6-phosphogluconate dehydrogenase, Ru5PE: ribulose-5-phosphate 3-epimerase, XI: xylose isomerase, XK: xylulose kinase.

The gaseous products (H_2 and CO_2) can be easily separated from the aqueous reactants under mild reaction conditions and the fast removal of the gaseous products drives the overall reaction in the direction of hydrogen formation. Another amazing feature of this endothermic reaction is that the chemical energy efficiency (output/input) in terms of higher heating values is 122%, suggesting that these systems can absorb low-temperature waste heat from the environment and then convert it to hydrogen energy.

Sugar-powered enzymatic fuel cells (EFC) are another notable example. The EFC is a bioelectro-chemical system that utilizes enzymes to convert chemical energy in fuels to electricity under mild conditions.[57,58] *Via* a synthetic enzymatic pathway, nearly 24 electrons are generated from one glucose unit of maltodextrin in an EFC.[42] With the complete oxidation of the sugar, such EFCs powered by a 15% (w/v) maltodextrin solution have an energy storage density of 596 A h kg^{-1}, which is one order of magnitude higher than that of lithium-ion batteries.[42]

High-yield production of ethanol[59] and *n*-butanol[38] from glucose has been demonstrated *via in vitro* reconstituted pathways. Another recent example is high-yield isoprene production *via* an *in vitro* mevalonate pathway with balancing of the ATP, NADPH and acetyl-CoA cofactors, resulting in ~100% molar yield.[60]

4.2.2   Fast Reaction Rate

Reaction rates per unit area or volume of enzymatic systems are usually faster than those based on whole cell systems because of the higher loadings of desired enzymes per unit area or volume and no cellular membranes, which are often rate-limiting for substrate assimilation or product secretion (Figure 4.3). For example, it is believed that EFCs will become the next generation of environmentally friendly micropower sources but not microbial fuel cells

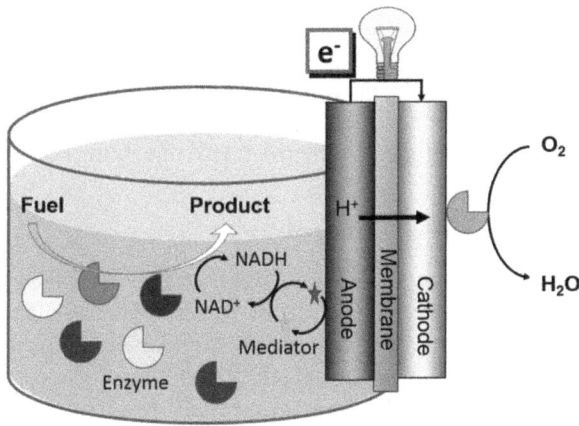

Figure 4.3   Enzymatic fuel cell *via* an enzymatic pathway for producing electricity.[42]

(MFCs),[61] because the former can exhibit far higher power density than the latter. For example, Logan[62] predicted that MFCs should have a maximum power density of 1.7 mW cm^{-2} at the anode based on their maximum substrate utilization rate. On the contrary, a number of EFC studies have demonstrated power densities at the milliwatt cm^{-2} level at the anode. For example, using a cascade of dehydrogenases with enzyme immobilization a polymer matrix, methanol and ethanol have been oxidized efficiently, generating power densities of 1.55 mW cm^{-2} and 2.04 mW cm^{-2}, respectively.[63] A glucose/O_2 EFC integrating carbon nanotube-based 3D enzyme-electrodes exhibits a maximum power output of 1.54 mW cm^{-2}.[64] By optimizing the EFC configuration, reaction buffer, membrane, and electron mediators, Sony has increased the maximum power densities of a sugar-powered EFC up to 5.0–10 mW cm^{-2}.[65,66]

Steroid hydroxylation reactions mediated by a cell extract containing P450 monooxygenase CYP106A2-catalyzed were 18-fold faster than in whole cells.[67] Cell-free protein synthesis has been widely used to synthesize and purify numerous proteins within several hours, more rapidly than microbial fermentations, which usually last several days.[47]

4.2.3 Easy Product Separation

Another merit of *in vitro* biosystems is easy and low-cost product separation from aqueous reaction solutions, especially for products that are usually located intracellularly, such as cytoplasmic proteins and sugar phosphates.[54] The recycling of biocatalysts is also important as it can save process costs and separate biocatalysts from the product easily. Enzyme immobilization has been widely used to recycle immobilized enzymes from the liquid reaction system.[68] For example, easily recyclable cellulose-containing magnetic nanoparticles have been developed for immobilizing enzymes.[69] These nanoparticles are new solid supports for immobilizing enzymes, which can be selectively recycled or removed from other biocatalysts, substrates, and products by a magnetic force.

In many cases, product separation could take place easily due to phase changes. In the cases of hydrogen production from sugars and water (Figure 4.2), gaseous hydrogen and carbon dioxide can bubble up from the aqueous solution.[36] In the case of starch made from cellulosic materials, long-chain linear starch can be precipitated in the presence of ethanol or by pH change.[44]

4.2.4 Easy Process and Control

Cell-free protein synthesis systems have been widely used for the production of *in vivo* cytotoxic, regulatory, or unstable proteins that are difficult to express in living cells. In principle, the biochemical environment for protein expression can be easily manipulated in cell-free systems. For example enzymes, with ecological and industrial importance include nitrogenases, methanogenic enzymes, cytochrome P450s, and hydrogenases.[70] [FeFe] hydrogenases are the fastest hydrogen-generation catalysts in many microbial ecosystems

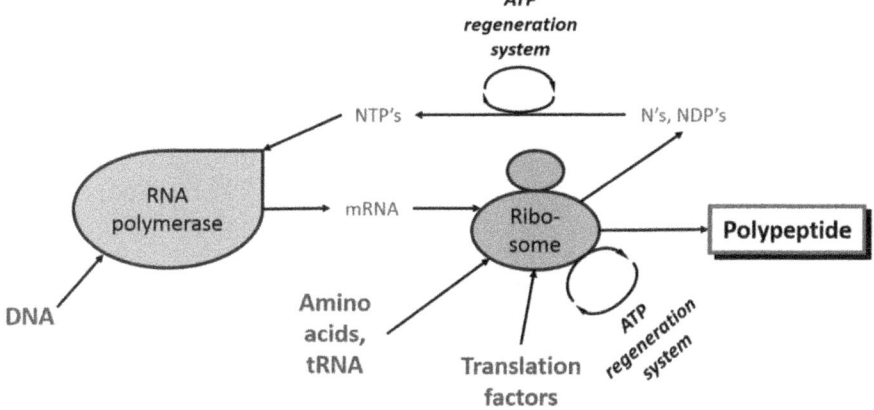

Figure 4.4   Scheme of cell-free protein synthesis.

and are also of great importance for producing hydrogen as a sustainable fuel. However, the recombinant expression of [FeFe] hydrogenases in living cells is challenging because such enzymes are extremely sensitive to oxygen; it requires the co-expression of heterologous helper proteins; and the experimental control over the expression and folding environment is limited by the barrier imposed by the cell wall. In a cell-free system, the absence of cell walls offers direct access to the protein production and folding environment, allowing facile addition of substrates, even to concentrations unachievable in living cells. The building bricks including DNA templates, amino acids, translation factors, mRNAs, nucleoside triphosphates, nucleoside diphosphates, the processing factories including RNA polymerase and ribosomes, along with the energy sources (the ATP regeneration system), are assembled in one pot (Figure 4.4).[12,71] Cell-free systems have been demonstrated to be consistently capable of expressing and maturing [FeFe] hydrogenases at much higher levels than have been achieved using whole cell systems.

Cell-free biosystems can assemble pathways easily using enzymes originated from different sources. These enzymes as building blocks are highly exchangeable. Furthermore, multi-building blocks can be combined to build modules with defined functions such as NADH regeneration and ATP generation.[30] For example, thermo-tolerant enzymes from different thermopiles have been assembled to construct artificial metabolic pathways *in vitro*, enabling the biotransformation of malate[72] or *n*-butanol[38] from glucose. These pathways and modules are easier to optimize and control than those *in vivo*.

4.2.5   Tolerance of Toxic Compounds

In vitro biosystems show better tolerance to toxic compounds than microorganisms because most organic solvents (*e.g.*, alcohols), ionic liquids, or antibiotics, disrupt cellular membranes to impair the basic metabolism of microorganisms.[73] For example, in an isobutanol fermentation by recombinant *E. coli*, concentrations of isobutanol as low as 1–2% (v/v) can induce toxic effects to the host, resulting in low product titers.[5,74] To solve this problem, an *in vitro* synthetic enzymatic pathway has been implemented to produce isobutanol (Figure 4.5).[39] The pathway scheme includes two parts: (1) the conversion of the glucose substrate to pyruvate; and (2) the generation of isobutanol from pyruvate. In part one, pyruvate is generated by four enzymes instead of the glycolysis pathway, resulting in no CoA being needed, different from the glycolysis pathway. This cell-free system is able to tolerate up to 4% isobutanol.[39]

In vitro biosystems also exhibit better tolerance to toxic compounds generated from biomass pretreatment. Dilute acid biomass hydrolysate contains a number of microorganism-toxic compounds, such as acetic acid and furfural, so that most ethanol-producing microorganisms cannot grow in the hydrolysate. A synthetic enzymatic pathway has been validated to achieve a very high yield conversion from xylose to xylitol powered by cellobiose.[43] This high-yield and projected low-lost biohydrogenation through cell-free enzymatic systems exhibited a great tolerance to the toxic biomass hydrolysate substrate and could be important for the production of sulfur-free liquid jet fuels in the future.[43]

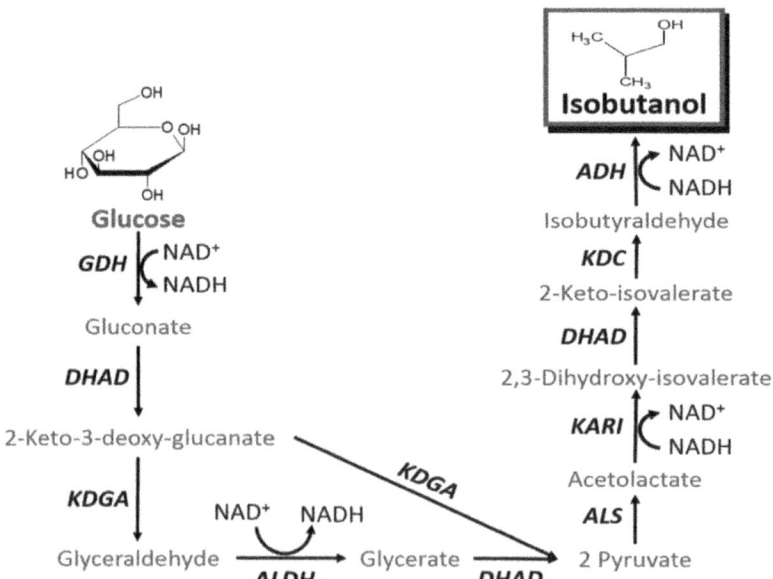

Figure 4.5 Production of isobutanol from glucose *via* an *in vitro* enzymatic pathway. GDH: glucose dehydrogenase; DHAD: (gluconate/glycerate) dihydroxy acid dehydratase; ALDH: glyceraldehyde dehydrogenase; KDGA: 2-keto-3-desoxygluconate aldolase; ALS: acetolactate synthase; KARI: ketolacid reductoisomerase; KDC: 2-ketoacid decarboxylase; PDC: pyruvate decarboxylase complex; ADH: alcohol dehydrogenase.

4.2.6 Capability of Performing Unnatural Reactions

In vitro biosystems have the capability of performing unnatural reactions that living cells or chemical catalysts cannot implement. For example, an enzyme cocktail has been demonstrated to convert pretreated biomass into edible starch, including endoglucanase, cellobiohydrolyase, cellobiose phosphorylase, and α-glucan phosphorylase originating from bacterial, fungal, and plant sources (Figure 4.6).[44] Although all of these enzymes are available in nature, cell membranes separate hydrolytic enzymes (*i.e.*, endoglucanase and cellobiohydrolase) from cytoplasmic enzymes (*i.e.*, cellobiose phosphorylase and α-glucan phosphorylase). As a result, this reaction cannot occur in nature. In this bioprocess, an ethanol-producing yeast is added to convert the remaining glucose to ethanol, a co-product. Yeast cells utilize the glucose to produce ethanol but not starch, glucose-1-phosphate, and cellobiose, dragging the overall reaction into the direction of amylose synthesis. Such co-production of amylose and ethanol does not waste any valuable sugars in biomass, indicating its great commercialization potential. This is a perfect example to demonstrate bioconversion through the combination of enzymes and living whole cells. In fact, *in vitro* biosystems do not strictly exclude the use of living cells, as long as it can bring in a positive outcome. Cellulose is the major component of plant cell walls, while starch is an important food source. The abundance of the former resource is approximately 50 times that of the latter and the cultivation of perennial plants has a far smaller environmental footprint than that of annual grain crops. This breakthrough from cellulose to starch implies an out-of-the-box solution to feed the world by the utilization of the most abundant non-food biomass resource.

4.2.7 Shifting Reaction Equilibrium

In vitro synthetic enzymatic pathways can be designed carefully to shift equilibrium intermediates to final products. The commercial process of fructose production from glucose comprises sequential reactions. The last step, an isoenergetic reaction, is the conversion of glucose to fructose *via* glucose isomerase, resulting in an equilibrium constant close to 45/55.[75] To increase the fructose percentage in the final product, a novel *in vitro* process has been

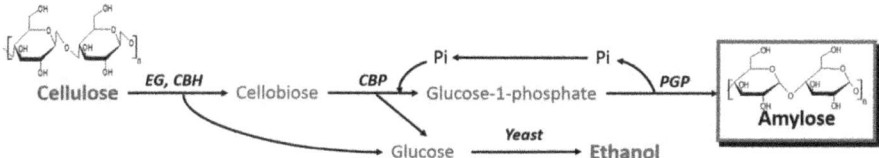

Figure 4.6 Coproduction of synthetic amylose and ethanol from cellulose *via* multi-enzymes and an ethanol-producing yeast. EG: endoglucanase, CBH: cellobiohydrolase, CBP: cellobiose phosphorylase, PGP: potato α-glucan phosphorylase.

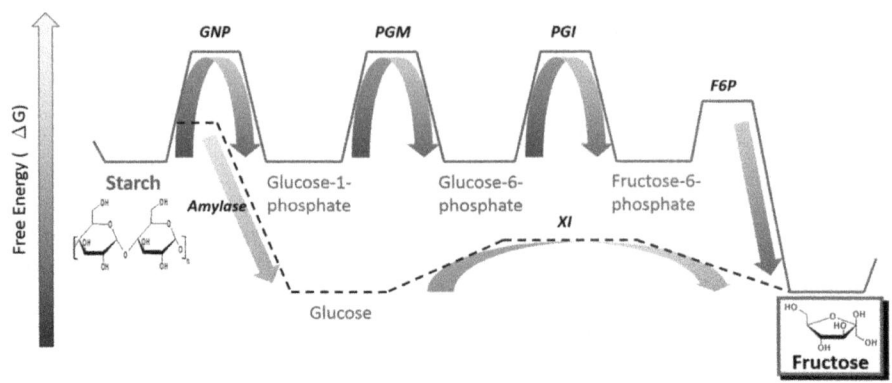

Figure 4.7   Two pathways for the production of fructose from starch. The dashed
line represents the traditional way. The solid line represents the new
pathway featuring a high product yield. GNP: glucan phosphorylase,
PGM: phosphoglucomutase, PGI: phosphoglucose isomerase, F6P:
fructose-6-phosphatase, XI: xylose isomerase.

designed to achieve a high percentage of fructose[45] (Figure 4.7). The four-
enzyme cascade includes starch phosphorylase, phosphoglucomutase, phosphoglucose isomerase, and fructose-6-phosphatase. The first three steps are
reversible and the final step is exergonic so that it can push the reaction
towards the product direction, resulting in a very high fructose yield. This
work illustrates general ideas that may prove useful in designing cell-free
synthetic enzymatic systems for other processes, in particular, the focus on
the energetics of the pathway.[45]

Similarly, the choice of enzymes in pathway design is of importance. In the
case of the conversion of cellulose to starch,[44] cellobiose phosphorylase and
α-glucan phosphorylase are responsible for reversibly converting from cellobiose to amylose or *vice versa*. It was found that α-glucan phosphorylase from
potato is a key enzyme to drive the reaction toward starch synthesis.[44] In contrast, the same enzyme from *Clostridium thermocellum* cannot generate amylase from cellobiose because it prefers the starch degradation direction.[44]

4.3   Biomimetic Cofactors: Role of Chemicals

A cofactor is a non-protein chemical compound that is required for an
enzyme's biological activity. It can be considered as a "helper molecule" that
assists in biochemical transformations. Cofactors can be classified into two
groups: organic cofactors, such as nicotinamide adenine dinucleotide phosphate (NADP), nicotinamide adenine dinucleotide (NAD), flavin adenine
dinucleotide (FAD), adenosine triphosphate (ATP), quinone compounds, and
coenzyme A (CoA), and inorganic cofactors, such as metal ions and iron–
sulfur clusters.[76] Because some organic cofactors are too costly to refurnish
by frequent addition, it is vital to regenerate them *in vitro*. NADPH from NADP
can be regenerated with various chemical, enzymatic, and electrochemical

methodologies.[77,78] Among them, enzymatic regeneration is the most popular for industrial application. On the other hand, some organic cofactors are not stable enough for repeated recycling. The most promising solution is the replacement of organic cofactors by using low-cost and stable biomimetic ones.[31,79] Such biomimetic cofactors are analogs that share a similar structure and function with the naturally occurring counterparts and can be chemically synthesized at a low price and with increased stability. Since natural NADP regeneration has been reviewed elsewhere,[21,78] here we highlight recent advances in biomimetic cofactors.

4.3.1   Biomimetic NADP

NADP consists of two nucleotides joined by a pair of bridging phosphate groups. The nucleotides have ribose rings, one with adenine attached to the first carbon atom position and the other with nicotinamide at this position (Figure 4.8). NADP has one extra phosphate group attached to the adenosine ribose compared to NAD. They play an essential role in many energy-related biochemical metabolisms, such as respiration and photosynthesis, because most oxidoreductase enzymes use NADP to transfer electrons and protons. NADP is usually used in anabolism,[80] while NAD is used in catabolism.[81] Such oxidation reactions involve the removal of two hydrogen atoms from the reactant, in the form of a hydride ion, and a proton. The oxidized NADP is reduced to NADH by the transfer of the hydride to the carbon opposite to the positive nitrogen in the nicotinamide ring, while the proton is released in solution.[82]

NAD is usually produced by microbial transformation *via* the salvage pathway for the biosynthesis of NAD from nicotinamide and ATP.[83] NADP can be produced by enzymatic phosphorylation of NAD. Their reduced forms, NADH and NADPH, can be obtained chemically or enzymatically from the oxidized forms. The production cost of industrial NAD in bulk supply is ~$1500 per kg, and the price for the reduced form is several-fold higher. In addition to their high price tags, the instability of NADPH further prohibits their application

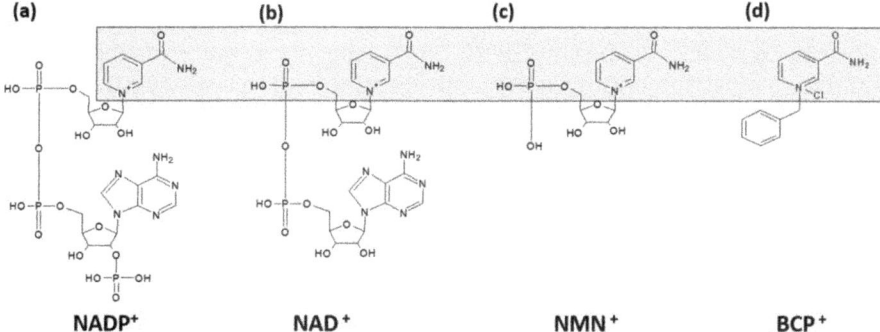

Figure 4.8   Chemical structures of (a) NADP, (b) NAD, and their biomimetics (c) NMN and (d) BCP.

in the large-scale production of biocommodities. NAD solutions are only stable for weeks at 4 °C and neutral pH, and decompose rapidly in acids and alkalis, or at room temperature. Upon decomposition, the degraded products are often enzyme inhibitors.[84]

Because NADP is less stable and more expensive than NAD, several methods have been used to engineer enzymes to switch cofactor preference from NADP to NAD.[85–88] For example, by using molecular modeling and comparing amino acid sequences responsible for cofactor binding sites, a NADP-preferring glutathione dehydrogenase was changed to a NAD-preferring one by site-directed mutagenesis.[89] More recently, NADP-preferring ketol-acid reductoisomerase and alcohol dehydrogenase were mutated to NAD-dependent ones, facilitating the balanced cofactors to achieve anaerobic high-yield iso-butanol fermentation.[90] A more general approach was established to completely reverse ketol-acid reductoisomerase cofactor dependence from NADP to NAD based on available crystal structure of this enzyme and a comprehensive protein sequence alignment.[91]

Because the pyrophosphate and adenosine groups associated with NADP are not essential in the hydride transfer, several small-size chemical compounds have been suggested to replace NADP.[82,92] Such biomimetic cofactors are less costly, have improved stability, and have fast mass transfer due to their reduced size. Nicotinamide mononucleotide (NMN), an electroactive half of the natural NAD, is considered a good alternative to NADP (Figure 4.8c). NMN retains its essentially functional nicotinamide ribose ring. It has been reported that several wild-type redox enzymes can work with this simpler version of NAD, including liver alcohol dehydrogenase[82] and glutamic dehydrogenase.[93] However, the observed activities towards NMN are extremely low. Recently, Banta and his coworkers demonstrated that an engineered *Pyrococcus furiosus* alcohol dehydrogenase has an ability to work with NMN used in enzymatic fuel cells.[94] Significant gains in fuel cell performance have been observed with NMN in immobilized systems due to its fast transport rate and small size, despite the decreased turnover rate of the enzyme.

Another low-cost and stable biomimetic cofactor, 1-benzyl-3-carbamoyl-pyridinium chloride (often called BCP), was suggested to replace natural cofactors by Fish.[82] Compared with NMN, the phosphate and adenosine groups are substituted by a more stable benzene ring in the structure of BCP (Figure 4.8d). Later, his group showed that wild-type horse liver alcohol dehydrogenase[93] and monooxygenase[95] can work on this biomimetic cofactor, although their activities are very weak. Another report shows that a wild-type enoate reductase can utilize BCP and its derivatives without compromising the activity or stereoselectivity of the bioreduction process.[96] Moreover, Clark and his coworkers demonstrated that engineered cytochrome P450 enzymes with two amino acids mutated can work on BCP at rates comparable to that of the natural cofactor.[97] Our lab has also succeeded in engineering NADP-dependent glucose 6-phosphate dehydrogenase and 6-phosphogluconate dehydrogenase to be BCP-dependent (in preparation for publication).

Engineering redox enzymes for biomimetic cofactors remains in its early stage because of a lack of a general approach that can deal with numerous enzymes with different and complicated catalytic mechanisms.[98–100]

4.3.2   Biomimetic Flavin

Flavin is a group of organic compounds based on pteridine formed by the tricyclic heteronuclear organic ring isoalloxazine (Figure 4.9a). The flavin moiety often forms as flavin mononucleotide (FMN) or flavin adenine dinucleotide (FAD). It is in one or the other of these forms that flavin is present as a prosthetic group on flavoproteins. As a cofactor, flavin is distinguished by its ability to catalyze a variety of different processes involving oxidation–reduction reactions.[101]

Riboflavin, also known as vitamin B_2, is the central component of FAD and FMN, and is therefore required by all flavo proteins. It plays an important role in the metabolism of fats, ketone bodies, carbohydrates, and proteins. Riboflavin for industrial use is mainly produced from ascomycete fungi in aerobic fermentation. Three quarters of riboflavin is used as a feed additive and the remaining is used as food additives and in pharmaceuticals.[102] Furthermore, FMN can be synthesized by chemical phosphorylation from riboflavin, while FAD can be produced by chemical synthesis or by microbial transformation, which uses FMN and ATP as the substrates.[83]

The aims of the modification of flavin-based compounds are to investigate the interaction of these flavin analogs with their reconstituted proteins and to understand their catalytic mechanism. Synthetic flavin analogs serve as active-site probes for almost all positions of the flavin nucleus.[102]

Flavin-derived coenzymes can perform a variety of redox reactions at the active sites of flavin-dependent enzymes. Outside of the enzymes, synthetic flavin analogs or biomimetic flavins have been used to catalyze numerous reactions with high selectivity and efficiency.[103,104] For example, Murray *et al.* used a biomimetic bridged flavin catalyst (Figure 4.9b) to implement the oxidation of alkyl and aryl aldehydes to their corresponding carboxylic acids.[105] This green, sustainable, and easily operated reaction uses low loadings of hydrogen peroxide and obviates chlorinated solvents with minimal

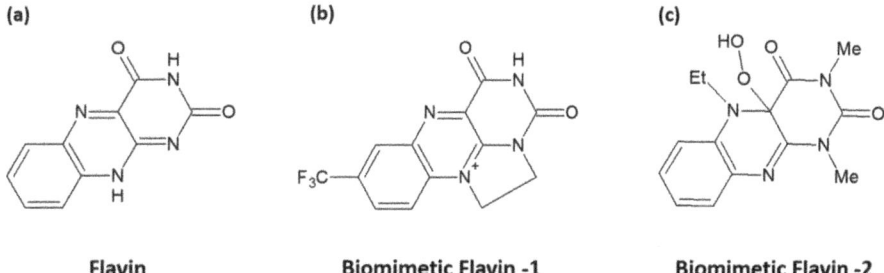

(a) Flavin   (b) Biomimetic Flavin -1   (c) Biomimetic Flavin -2

Figure 4.9   Chemical structures of (a) flavin and (b) and (c) its biomimetics.

byproducts. In another example, selective *cis*-dihydroxylation of olefins with the aid of a biomimetic flavin-based coupled catalytic system (Figure 4.9c) using hydrogen peroxide as the terminal oxidant has been developed.[106] The present biomimetic catalytic system works well in asymmetric dihydroxylation and gives optically active diols in good isolated yields and high enantiomeric excesses. Examples include: the organocatalytic Dakin oxidation of electron-rich aryladehyde to phenols using biomimetic flavin catalysts under mild, basic conditions;[107] the efficient and selective sulfoxidation by hydrogen peroxide using a recyclable biomimetic flavin plus ionic liquid catalytic system;[108] and the sulfoxidation of methyl phenyl sulfides with hydrogen peroxide using β-cyclodextrin–flavin conjugates as highly efficient catalysts.[109]

4.3.3   Biomimetic ATP (Polyphosphate)

ATP, a nucleoside triphosphate, is an energy currency used in cells and is a coenzyme for enzymes in transduction mechanisms, in metabolic pathways, and in the regulation of enzyme, channel, and receptor activities. One ATP molecule contains three phosphate groups attached to an adenosine group (Figure 4.10a). The two phosphoanhydride bonds are responsible for the high energy content of the ATP.[110] Natural ATP is often produced from yeast with a relatively high production cost. It is stable in solution at neutral pH, but is rapidly hydrolyzed at extreme pHs and is unstable in unbuffered water. Numerous ATP analogues have been synthesized to probe the role of ATP in biosystems. A good review has summarized a representative selection of these derivatives with their key properties and functions.[111] Three kinds of ATP analogues can be synthesized based on the position of the modification, which occurs on the triphosphate group, the base, or the ribose. The ease of synthesising them varies widely depending on skills, facilities, and reaction conditions.

Besides these ATP analogues, polyphosphate is a linear polymer of tens or hundreds of phosphate residues (Figure 4.10b). In the pre-ATP world, this ubiquitous and abundant biopolymer is hypothesized to have played a vital

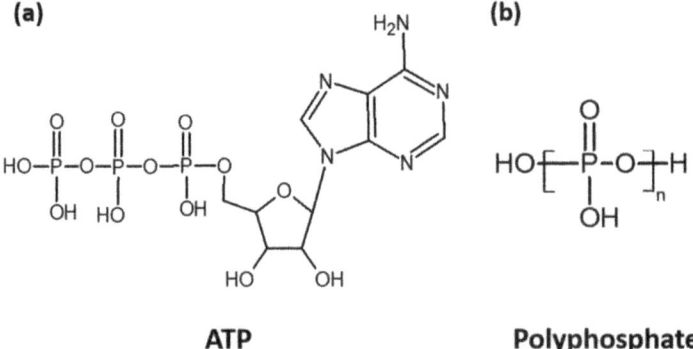

Figure 4.10   Chemical structures of (a) ATP and (b) polyphosphate.

role in the energy metabolism and as a stored energy source. Polyphosphate can be produced by dehydration of phosphoric acid at an elevated temperature or through biological synthesis by polyphosphate-accumulating microorganisms.[112] Due to its lower cost and higher stability compared to other phosphoryl donor compounds, such as acetyl phosphate, phosphoenol pyruvate, and creatine phosphate, polyphosphate is used for ATP generation/regeneration systems in preparative-scale organic synthesis and biocatalysis.[113,114] These ATP regeneration systems usually involve polyphosphate kinase and phosphortransferase.[115]

Polyphosphate can be used in *in vitro* biosystems as an inexpensive and stable phosphoryl donor. For instance, Murata *et al.* succeeded in conferring the ability of γ-proteobacterial ATP-specific NAD kinases to be polyphosphate-dependent through a single amino-acid substitution.[116] The group also demonstrated that this created polyphosphate/ATP-NADK was suitable for the polyphosphate-dependent mass production of NADP from NAD. In another example, Zhang *et al.* produce high-yield hydrogen from xylose *via* a synthetic enzymatic pathway containing a xylulo kinase that utilizes polyphosphate instead of costly ATP.[35]

4.4   Perspectives

In vitro biosystems for biomanufacturing are still at the development stage. In addition to costly and labile cofactors, several challenges are hampering their scale-up for cost-effective biomanufacturing, including high enzyme costs, poor enzyme stability, different optimal conditions for enzymes, and optimization of enzymatic pathways (Table 4.1).

For the production of low-selling price biocommodities, *in vitro* biosystems require low enzyme production costs. Unlike high-purity enzymes, antibodies, and glycosylated proteins as drugs, most enzymes used in *in vitro* biosystems do not need high purity or complicated post-translation modifications; even cell extracts work. According to well-known industrial enzyme production, several solutions are the choice of hosts able to highly secrete enzymes,[4] high-level intracellular protein expression hosts,[117] high-cell-density fermentation.[118] Simple enzyme purification techniques include heat precipitation,[119] ammonia sulfate precipitation,[120] simple adsorption and immobilization,[121] resin-free chromatographic separation,[122] and multienzyme co-immobilization and co-purification.[123] Several industrial enzymes such as protease and amylase produced by *Bacillus* sp., cellulase produced by *Trichoderma* and *Aspergillus* sp., have been sold at $5–10 per kg of protein.[4,124] It is estimated that the current cost of recombinant proteins produced by *E. coli* BL 21 can be reduced to ~$100 per kg of protein or lower.[125,126] Like most industrial processes, production costs will decrease rapidly when each part is improved and the whole system is integrated.

Low enzyme stability may be another limitation of *in vitro* biosystems. The total turnover number (TTN) of the enzyme represents the lifetime of the biocatalyst under its working conditions. Enzyme stability can be addressed

Table 4.1 Possible problems of *in vitro* biosystems comprised of synthetic enzymatic pathways and their respective solutions.

Problems	Solutions
Costly and labile cofactors	Cofactor immobilization
	Cofactor regeneration
	Replacement of stable and low-cost biomimetic cofactors
High enzyme cost	Secretory enzymes
	High-expression host
	High-cell-density fermentation
	Simple enzyme purification techniques
	• Heat precipitation
	• Ammonia sulfate precipitation
	• Simple adsorption and immobilization
	• Resin-free chromatographic separation
	Multi-enzyme co-immobilization and co-purification
Poor enzyme stability	Use of thermostable enzymes
	Enzyme immobilization
	Use of engineered enzymes
	• Rational design
	• Directed evolution
	• Combinatorial strategy
Different optimal conditions for enzymes	Discovery of enzymes from one source
	Use of engineered enzymes
	Adjusted reaction conditions in terms of time
Optimization of enzymatic pathways	Modulation and localization of enzymes
	Mathematical modeling
	High-throughput technology

by using thermostable enzymes derived from thermophilic hosts, enzyme immobilization, and enzyme engineering through rational design, directed evolution, or a combinatorial strategy. Many thermostable enzymes have been discovered from numerous thermophilic microorganisms, such as *Clostridium thermocellum* and *Thermotoga maritima*.[119,127,128] Enzyme immobilization has been demonstrated to stabilize enzymes for a long time. In the past half century, many enzyme immobilization technologies have been developed, including non-covalent adsorption and deposition, ionic interaction adsorption, covalent attachment, crosslinking, and entrapment in a polymeric gel or capsule.[30,129] Besides enzyme immobilization, enzymes can be engineered to increase their stability. Directed evolution generates random mutants, then among them, the most stable ones are screened.[52,130] Rational design referred to mutations are designed and predicted based on the known information about the enzyme.[131] A combination of rational design and directed evolution could save time to create more stable enzymes within a short time.[132] Numerous enzymes have been developed to meet industrial needs with high stability.[2,133]

As several enzymes or enzymatic pathways are put in one pot, these enzymes may have to suffer from compromised activities at non-optimal conditions.[36] For example, at room temperature, some enzymes originated from hyperthermophilic microorganisms exhibit very low activities.[37] However, this potential problem can be solved by discovering more enzymes from the same source, using engineered enzymes with improved catalytic performances, and adjusting reaction conditions in the course of the reaction.[31,134]

Another challenge relates to the optimization of the pathways. Although the enzymatic pathway in cell-free systems is much simpler than in living cell systems, the elucidation and manipulation of the enzymatic pathway to obtain the optimal product outcome is still difficult,[10,135,136] but much easier than in *in vivo* systems.[137,138] One approach of modulating and localizing several cascade enzymes to assemble an enzyme complex has been successful in understanding enzyme cascade reactions.[133,139,140] Mathematical modelling is another powerful tool to optimize multi-enzyme reactions.[141–143] For example, Zeng and his coworker applied a genetic algorithm to study and predict the yield of the biohydrogen production in an *in vitro* biosystem.[144] Furthermore, high-throughput techniques such as biochips and micro-reactors have emerged as a state-of-the-art tool for pathway construction and optimization.[145,146]

4.5 Conclusions

The concept of using *in vitro* biosystems to produce high-value specialty chemicals, fine chemicals, and biopharmaceuticals is becoming a reality. This concept is being expanded to produce low-value biocommodities, including hydrogen, electricity, liquid biofuels, food, and feed, featuring high product yields, fast reaction rates, high tolerance of toxic compounds, capability of performing unnatural reactions, the ability to shift the reaction equilibrium, and easy process control and regulation. Designing and utilizing biomimetic cofactors, such as biomimetic NADP, flavin, and ATP, holds great potential for decreasing manufacturing costs and expanding the potential of biocatalysis to a lot of complicated products rather than hydrolysis or isomerization mediated by cofactor-free enzymes. Several obstacles regarding enzyme production cost, enzyme stability, compromised enzyme activity, and the optimization of enzymatic pathways will be addressed within a short time when more and more companies realize the great potential of *in vitro* biosystems. With more and more effective and stable enzymes and cofactors, *in vitro* biosystems will become a widely used industrial biotransformation platform, in particular, for the production of biofuels, value-added biochemicals, and food/feed.

Acknowledgments

The authors thank the Biological Systems Engineering Department of Virginia Tech for their support. This material is mainly based upon work supported by the National Science Foundation grant no. (IIP-1214895) to PZ and Department of Energy grant to ZZ.

References

1. U. T. Bornscheuer, G. W. Huisman, R. J. Kazlauskas, S. Lutz, J. C. Moore and K. Robins, *Nature*, 2012, **485**, 185–194.
2. S. Panke, M. Held and M. Wubbolts, *Curr. Opin. Biotechnol.*, 2004, **15**, 272–279.
3. P. A. Santacoloma, G. Sin, K. V. Gernaey and J. M. Woodley, *Org. Process Res. Dev.*, 2010, **15**, 203–212.
4. Y. H. P. Zhang, *Biotechnol. Bioeng.*, 2010, **105**, 663–677.
5. S. Atsumi, T. Hanai and J. C. Liao, *Nature*, 2008, **451**, 86–U13.
6. A. Zaks, *Curr. Opin. Chem. Biol.*, 2001, **5**, 130–136.
7. E. J. Steen, Y. Kang, G. Bokinsky, Z. Hu, A. Schirmer, A. McClure, S. B. del Cardayre and J. D. Keasling, *Nature*, 2010, **463**, 559–562.
8. P. P. Ng, M. Jia, K. G. Patel, J. D. Brody, J. R. Swartz, S. Levy and R. Levy, *Proc. Natl. Acad. Sci. U. S. A.*, 2012, **109**, 14526–14531.
9. M. C. Jewett, K. A. Calhoun, A. Voloshin, J. J. Wuu and J. R. Swartz, *Mol. Syst. Biol.*, 2008, **4**, 220.
10. W. C. Yang, K. G. Patel, H. E. Wong and J. R. Swartz, *Biotechnol. Prog.*, 2012, **28**, 413–420.
11. J. R. Swartz, *AIChE J.*, 2012, **58**, 5–13.
12. J. Swartz, *J. Ind. Microbiol. Biotechnol.*, 2006, **33**, 476–485.
13. J. R. Swartz, *Nat. Biotechnol.*, 2009, **27**, 731–732.
14. I. Pervaiz, S. Ahmad, M. A. Madni, H. Ahmad and F. H. Khaliq, *Appl. Biochem. Microbiol.*, 2013, **49**, 437–450.
15. D. Vasic-Racki, in *Industrial Biotransformations*, Wiley-VCH Verlag GmbH & Co. KGaA, 2006, pp. 1–36.
16. J.-K. Guterl and V. Sieber, *Eng. Life Sci.*, 2013, **13**, 4–18.
17. Y. H. P. Zhang, J. B. Sun and J. J. Zhong, *Curr. Opin. Biotechnol.*, 2010, **21**, 663–669.
18. E. Ricca, B. Brucher and J. H. Schrittwieser, *Adv. Synth. Catal.*, 2011, **353**, 2239–2262.
19. R. Wolfenden and M. J. Snider, *Acc. Chem. Res.*, 2001, **34**, 938–945.
20. E. Buchner, *Ber. Dtsch. Chem. Ges.*, 1897, **30**, 117–124.
21. R. Wichmann and D. Vasic-Racki, *Adv. Biochem. Eng./Biotechnol.*, 2005, **92**, 225–260.
22. Y.-H. P. Zhang and L. R. Lynd, *Biotechnol. Bioeng.*, 2004, **88**, 797–824.
23. A. M. Daines, B. A. Maltman and S. L. Flitsch, *Curr. Opin. Chem. Biol.*, 2004, **8**, 106–113.
24. W. D. Fessner and V. Helaine, *Curr. Opin. Biotechnol.*, 2001, **12**, 574–586.
25. H. Malekan, G. Fung, V. Thon, Z. Khedri, H. Yu, J. Y. Qu, Y. H. Li, L. Ding, K. S. Lam and X. Chen, *Bioorg. Med. Chem.*, 2013, **21**, 4778–4785.
26. M. M. Muthana, J. Qu, Y. Li, L. Zhang, H. Yu, L. Ding, H. Malekan and X. Chen, *Chem. Commun.*, 2012, **48**, 2728–2730.
27. M. Pesic, C. Lopez, J. Lopez-Santin and G. Alvaro, *Appl. Microbiol. Biotechnol.*, 2013, **97**, 7173–7183.
28. C. Hold and S. Panke, *J. R. Soc., Interface*, 2009, **6**, S507–S521.

29. A. C. Forster and G. M. Church, *Genome Res.*, 2007, **17**, 1–6.
30. Y. H. P. Zhang, S. Myung, C. You, Z. G. Zhu and J. A. Rollin, *J. Mater. Chem.*, 2011, **21**, 18877–18886.
31. C. You and Y.-H. P. Zhang, *Adv. Biochem. Eng./Biotechnol.*, 2013, **131**, 89–119.
32. Y. Shimizu, A. Inoue, Y. Tomari, T. Suzuki, T. Yokogawa, K. Nishikawa and T. Ueda, *Nat. Biotechnol.*, 2001, **19**, 751–755.
33. K. A. Thiel, *Nat. Biotechnol.*, 2004, **22**, 1365–1372.
34. Y. Wang and Y.-H. P. Zhang, *BMC Biotechnol.*, 2009, **9**, 58.
35. J. S. Martín del Campo, J. Rollin, S. Myung, Y. Chun, S. Chandrayan, R. Patiño, M. W. W. Adams and Y.-H. P. Zhang, *Angew. Chem., Int. Ed.*, 2013, **52**, 4587–4590.
36. Y.-H. P. Zhang, B. R. Evans, J. R. Mielenz, R. C. Hopkins and M. W. W. Adams, *PLoS One*, 2007, **2**, e456.
37. X. Ye, Y. Wang, R. C. Hopkins, M. W. W. Adams, B. R. Evans, J. R. Mielenz and Y.-H. P. Zhang, *ChemSusChem*, 2009, **2**, 149–152.
38. B. Krutsakorn, K. Honda, X. Ye, T. Imagawa, X. Bei, K. Okano and H. Ohtake, *Metab. Eng.*, 2013, **20**, 84–91.
39. J.-K. Guterl, D. Garbe, J. Carsten, F. Steffler, B. Sommer, S. Reiße, A. Philipp, M. Haack, B. Rühmann, A. Koltermann, U. Kettling, T. Brück and V. Sieber, *ChemSusChem*, 2012, **5**, 2165–2172.
40. M. J. Moehlenbrock and S. D. Minteer, *Chem. Soc. Rev.*, 2008, **37**, 1188–1196.
41. M. J. Cooney, V. Svoboda, C. Lau, G. Martin and S. D. Minteer, *Energy Environ. Sci.*, 2008, **1**, 320–337.
42. Z. G. Zhu, T. K. Tam, F. F. Sun, C. You and Y. H. P. Zhang, *Nat. Commun.*, 2014, **5**, 3026.
43. Y. Wang, W. Huang, N. Sathitsuksanoh, Z. Zhu and Y.-H. P. Zhang, *Chem. Biol.*, 2011, **18**, 372–380.
44. C. You, H. Chen, S. Myung, N. Sathitsuksanoh, H. Ma, X.-Z. Zhang, J. Li and Y.-H. P. Zhang, *Proc. Natl. Acad. Sci. U. S. A.*, 2013, **110**, 7182–7187.
45. A. Moradian and S. A. Benner, *J. Am. Chem. Soc.*, 1992, **114**, 6980–6987.
46. J. A. Rollin, W. Tam and Y. H. P. Zhang, *Green Chem.*, 2013, **15**, 1708–1719.
47. F. Katzen, G. Chang and W. Kudlicki, *Trends Biotechnol.*, 2005, **23**, 150–156.
48. Y. H. P. Zhang, *ACS Catal.*, 2011, **1**, 998–1009.
49. W. D. Huang and Y.-H. P. Zhang, *Energy Environ. Sci.*, 2011, **4**, 784–792.
50. Y.-H. P. Zhang, *Energy Sci. Eng.*, 2013, **1**, 27–41.
51. N. J. Turner, *Trends Biotechnol.*, 2003, **21**, 474–478.
52. V. G. H. Eijsink, S. Gaseidnes, T. V. Borchert and B. van den Burg, *Biomol. Eng.*, 2005, **22**, 21–30.
53. W. Tischer and V. Kasche, *Trends Biotechnol.*, 1999, **17**, 326–335.
54. Y.-H. P. Zhang and W.-D. Huang, *Trends Biotechnol.*, 2012, **30**, 301–306.
55. J. Woodward, M. Orr, K. Cordray and E. Greenbaum, *Nature*, 2000, **405**, 1014–1015.

56. R. K. Thauer, K. Jungermann and K. Decker, *Bacteriol. Rev.*, 1977, **41**, 100–180.
57. Z. G. Zhu, F. F. Sun, X. Z. Zhang and Y. H. P. Zhang, *Biosens. Bioelectron.*, 2012, **36**, 110–115.
58. F. Davis and S. P. J. Higson, *Biosens. Bioelectron.*, 2007, **22**, 1224–1235.
59. P. Welch and R. K. Scopes, *J. Biotechnol.*, 1985, **2**, 257–273.
60. T. P. Korman, B. Sahachartsiri, D. Li, J. M. Vinokur, D. Eisenberg and J. U. Bowie, *Protein Sci.*, 2014, **23**(5), 576–585.
61. M. H. Osman, A. A. Shah and F. C. Walsh, *Biosens. Bioelectron.*, 2011, **26**, 3087–3102.
62. B. E. Logan, *Nat. Rev. Microbiol.*, 2009, **7**, 375–381.
63. N. L. Akers, C. M. Moore and S. D. Minteer, *Electrochim. Acta*, 2005, **50**, 2521–2525.
64. B. Reuillard, A. Le Goff, C. Agnes, M. Holzinger, A. Zebda, C. Gondran, K. Elouarzaki and S. Cosnier, *Phys. Chem. Chem. Phys.*, 2013, **15**, 4892–4896.
65. Y. Tokita, T. Nakagawa, H. Sakai, T. Sugiyama, R. Matsumoto and T. Hatazawa, *ECS Trans.*, 2008, **13**, 89–97.
66. H. Sakai, T. Nakagawa, H. Mita, R. Matsumoto, T. Sugiyama, H. Kumita, Y. Tokita and T. Hatazawa, *ECS Trans.*, 2009, **16**, 9–15.
67. D. Zehentgruber, F. Hannemann, S. Bleif, R. Bernhardt and S. Lutz, *ChemBioChem*, 2010, **11**, 713–721.
68. R. C. Rodrigues, C. Ortiz, A. Berenguer-Murcia, R. Torres and R. Fernandez-Lafuente, *Chem. Soc. Rev.*, 2013, **42**, 6290–6307.
69. S. W. Myung, C. You and Y. H. P. Zhang, *J. Mater. Chem. B*, 2013, **1**, 4419–4427.
70. M. E. Boyer, J. A. Stapleton, J. M. Kuchenreuther, C. W. Wang and J. R. Swartz, *Biotechnol. Bioeng.*, 2008, **99**, 59–67.
71. T. W. Kim, J. W. Keum, I. S. Oh, C. Y. Choi, C. G. Park and D. M. Kim, *J. Biotechnol.*, 2006, **126**, 554–561.
72. X. Ye, K. Honda, T. Sakai, K. Okano, T. Omasa, R. Hirota, A. Kuroda and H. Ohtake, *Microb. Cell Fact.*, 2012, **11**, 120.
73. J. Zaldivar, J. Nielsen and L. Olsson, *Appl. Microbiol. Biotechnol.*, 2001, **56**, 17–34.
74. S. Atsumi, T. Y. Wu, I. M. P. Machado, W. C. Huang, P. Y. Chen, M. Pellegrini and J. C. Liao, *Mol. Syst. Biol.*, 2010, **6**, 449.
75. R. K. Bandlish, J. Michael Hess, K. L. Epting, C. Vieille and R. M. Kelly, *Biotechnol. Bioeng.*, 2002, **80**, 185–194.
76. K. A. Denessiouk, V.-V. Rantanen and M. S. Johnson, *Proteins: Struct., Funct., Bioinf.*, 2001, **44**, 282–291.
77. U. Kragl, W. Kruse, W. Hummel and C. Wandrey, *Biotechnol. Bioeng.*, 1996, **52**, 309–319.
78. H. Zhao and W. A. van der Donk, *Curr. Opin. Biotechnol.*, 2003, **14**, 583–589.
79. P. L. Hentall, N. Flowers and T. D. H. Bugg, *Chem. Commun.*, 2001, 2098–2099.

80. J. D. Horton, J. L. Goldstein and M. S. Brown, *J. Clin. Invest.*, 2002, **109**, 1125–1131.
81. C. Garrigues, P. Loubiere, N. D. Lindley and M. Cocaign-Bousquet, *J. Bacteriol.*, 1997, **179**, 5282–5287.
82. H. C. Lo, C. Leiva, O. Buriez, J. B. Kerr, M. M. Olmstead and R. H. Fish, *Inorg. Chem.*, 2001, **40**, 6705–6716.
83. S. Shimizu, in *Biotechnology*, Wiley-VCH Verlag GmbH, 2008, pp. 318–340.
84. N. Pollak, C. Dolle and M. Ziegler, *Biochem. J.*, 2007, **402**, 205–218.
85. S. Banta, B. A. Swanson, S. Wu, A. Jarnagin and S. Anderson, *Biochemistry*, 2002, **41**, 6226–6236.
86. S. Banta, B. A. Swanson, S. Wu, A. Jarnagin and S. Anderson, *Protein Eng.*, 2002, **15**, 131–140.
87. L. Zhang, B. Ahvazi, R. Szittner, A. Vrielink and E. Meighen, *Biochemistry*, 1999, **38**, 11440–11447.
88. A. Rosell, E. Valencia, W. F. Ochoa, I. Fita, X. Pares and J. Farres, *J. Biol. Chem.*, 2003, **278**, 40573–40580.
89. N. S. Scrutton, A. Berry and R. N. Perham, *Nature*, 1990, **343**, 38–43.
90. S. Bastian, X. Liu, J. T. Meyerowitz, C. D. Snow, M. M. Y. Chen and F. H. Arnold, *Metab. Eng.*, 2011, **13**, 345–352.
91. S. Brinkmann-Chen, T. Flock, J. K. B. Cahn, C. D. Snow, E. M. Brustad, J. A. McIntosh, P. Meinhold, L. Zhang and F. H. Arnold, *Proc. Natl. Acad. Sci. U. S. A.*, 2013, **110**, 10946–10951.
92. H. C. Lo, O. Buriez, J. B. Kerr and R. H. Fish, *Angew. Chem., Int. Ed.*, 1999, **38**, 1429–1432.
93. H. C. Lo and R. H. Fish, *Angew. Chem., Int. Ed.*, 2002, **41**, 478–481.
94. E. Campbell, M. Meredith, S. D. Minteer and S. Banta, *Chem. Commun.*, 2012, **48**, 1898–1900.
95. J. Lutz, F. Hollmann, T. V. Ho, A. Schnyder, R. H. Fish and A. Schmid, *J. Organomet. Chem.*, 2004, **689**, 4783–4790.
96. C. E. Paul, S. Gargiulo, D. J. Opperman, I. Lavandera, V. Gotor-Fernandez, V. Gotor, A. Taglieber, I. W. C. E. Arends and F. Hollmann, *Org. Lett.*, 2013, **15**, 180–183.
97. J. D. Ryan, R. H. Fish and D. S. Clark, *ChemBioChem*, 2008, **9**, 2579–2582.
98. E. Campbell, I. R. Wheeldon and S. Banta, *Biotechnol. Bioeng.*, 2010, **107**, 763–774.
99. Y. Wang, K.-Y. San and G. N. Bennett, *Curr. Opin. Biotechnol.*, 2013, **24**, 994–999.
100. M. J. Rane and K. C. Calvo, *Arch. Biochem. Biophys.*, 1997, **338**, 83–89.
101. L. Michaelis, M. P. Schubert and C. V. Smythe, *J. Biol. Chem.*, 1936, **116**, 587–607.
102. S. Chapman, G. Reid, D. Edmondson and S. Ghisla, in *Flavoprotein Protocols*, Humana Press, 1999, vol. 131, pp. 157–179.
103. S. Ghisla and V. Massey, *Biochem. J.*, 1986, **239**, 1–12.
104. Y. V. S. N. Murthy and V. Massey, *Vitamins and Coenzymes, Pt J*, 1997, vol. 280, pp. 436–460.

105. A. T. Murray, P. Matton, N. W. G. Fairhurst, M. P. John and D. R. Carbery, *Org. Lett.*, 2012, **14**, 3656–3659.
106. S. Y. Jonsson, K. Farnegardh and J. E. Backvall, *J. Am. Chem. Soc.*, 2001, **123**, 1365–1371.
107. S. Chen, M. S. Hossein and F. W. Foss, *Org. Lett.*, 2012, **14**, 2806–2809.
108. A. A. Linden, M. Johansson, N. Hermanns and J. E. Backvall, *J. Org. Chem.*, 2006, **71**, 3849–3853.
109. V. Mojr, V. Herzig, M. Budesinsky, R. Cibulka and T. Kraus, *Chem. Commun.*, 2010, **46**, 7599–7601.
110. J. R. Knowles, *Annu. Rev. Biochem.*, 1980, **49**, 877–919.
111. C. R. Bagshaw, *J. Cell Sci.*, 2001, **114**, 459–460.
112. A. Kuroda, K. Nomura, R. Ohtomo, J. Kato, T. Ikeda, N. Takiguchi, H. Ohtake and A. Kornberg, *Science*, 2001, **293**, 705–708.
113. S. M. Resnick and A. J. B. Zehnder, *Appl. Environ. Microbiol.*, 2000, **66**, 2045–2051.
114. A. Kameda, T. Shiba, Y. Kawazoe, Y. Satoh, Y. Ihara, M. Munekata, K. Ishige and T. Noguchi, *J. Biosci. Bioeng.*, 2001, **91**, 557–563.
115. M. Sato, Y. Masuda, K. Kirimura and K. Kino, *J. Biosci. Bioeng.*, 2007, **103**, 179–184.
116. Y. Nakamichi, A. Yoshioka, S. Kawai and K. Murata, *Sci. Rep.*, 2013, **3**, 2632.
117. S. C. Makrides, *Microbiol. Rev.*, 1996, **60**, 512–538.
118. J. Shiloach and R. Fass, *Biotechnol. Adv.*, 2005, **23**, 345–357.
119. Y. R. Wang and Y. H. P. Zhang, *Microb. Cell Fact.*, 2009, **8**, 30.
120. F. F. Sun, X. Z. Zhang, S. Myung and Y.-H. P. Zhang, *Protein Expression Purif.*, 2012, **82**, 302–307.
121. S. Myung, X. Z. Zhang and Y. H. P. Zhang, *Biotechnol. Prog.*, 2011, **27**, 969–975.
122. H. H. Liao, S. Myung and Y.-H. P. Zhang, *Appl. Microbiol. Biotechnol.*, 2012, **93**, 1109–1117.
123. C. You, S. Myung and Y. H. P. Zhang, *Angew. Chem., Int. Ed.*, 2012, **51**, 8787–8790.
124. D. Klein-Marcuschamer, P. Oleskowicz-Popiel, B. A. Simmons and H. W. Blanch, *Biotechnol. Bioeng.*, 2012, **109**, 1083–1087.
125. P. r. Tufvesson, J. Lima-Ramos, M. Nordblad and J. M. Woodley, *Org. Process Res. Dev.*, 2011, **15**, 266–274.
126. Z. Zhu, T. K. Tam and Y.-H. P. Zhang, *Adv. Biochem. Eng./Biotechnol.*, 2012, **131**, 89–119.
127. Y. Wang and Y. H. P. Zhang, *J. Appl. Microbiol.*, 2010, **108**, 39–46.
128. S. Myung, Y. R. Wang and Y.-H. P. Zhang, *Process Biochem.*, 2010, **45**, 1882–1887.
129. J. Kim, J. W. Grate and P. Wang, *Chem. Eng. Sci.*, 2006, **61**, 1017–1026.
130. J. D. Bloom and F. H. Arnold, *Proc. Natl. Acad. Sci. U. S. A.*, 2009, **106**, 9995–10000.
131. V. G. H. Eijsink, A. Bjork, S. Gaseidnes, R. Sirevag, B. Synstad, B. van den Burg and G. Vriend, *J. Biotechnol.*, 2004, **113**, 105–120.

132. X. Ye, C. Zhang and Y.-H. P. Zhang, *Mol. BioSyst.*, 2012, **8**, 1815–1823.
133. R. J. Conrado, J. D. Varner and M. P. DeLisa, *Curr. Opin. Biotechnol.*, 2008, **19**, 492–499.
134. R. Schoevaart, F. van Rantwijk and R. A. Sheldon, *Chem. Commun.*, 1999, 2465–2466.
135. P. Kar, H. Wen, H. Z. Li, S. D. Minteer and S. C. Barton, *J. Electrochem. Soc.*, 2011, **158**, B580–B586.
136. W. Johnston, N. Maynard, B. Y. Liaw and M. J. Cooney, *Enzyme Microb. Technol.*, 2006, **39**, 131–140.
137. X. Yu, T. Liu, F. Zhu and C. Khosla, *Proc. Natl. Acad. Sci. U. S. A.*, 2011, **108**, 18643–18648.
138. F. Zhu, X. Zhong, M. Hu, L. Lu, Z. Deng and T. Liu, *Biotechnol. Bioeng.*, 2014, **111**, 1396–1405.
139. M. J. Moehlenbrock, T. K. Toby, A. Waheed and S. D. Minteer, *J. Am. Chem. Soc.*, 2010, **132**, 6288–6289.
140. Y. H. P. Zhang, *Biotechnol. Adv.*, 2011, **29**, 715–725.
141. W. Johnston, M. J. Cooney, B. Y. Liaw, R. Sapra and M. W. W. Adams, *Enzyme Microb. Technol.*, 2005, **36**, 540–549.
142. J. Schmider, D. J. Greenblatt, J. S. Harmatz and R. I. Shader, *Br. J. Clin. Pharmacol.*, 1996, **41**, 593–604.
143. I. Ardao, E. Hwang and A.-P. Zeng, *Adv. Biochem. Eng./Biotechnol.*, 2013, **137**, 153–184.
144. I. Ardao and A.-P. Zeng, *Chem. Eng. Sci.*, 2013, **87**, 183–193.
145. D. Wahler and J.-L. Reymond, *Curr. Opin. Biotechnol.*, 2001, **12**, 535–544.
146. G. Y. Jung and G. Stephanopoulos, *Science*, 2004, **304**, 428–431.

CHAPTER 5

Chemical Bioengineering in Microbial Electrochemical Systems

DAN-DAN ZHAI[a,b] AND YANG-CHUN YONG*[a]

[a]Biofuels Institute, School of the Environment, Jiangsu University, Zhenjiang 212013, Jiangsu Province, China; [b]School of Bioengineering, Henan University of Technology, Zhengzhou, China
*E-mail: ycyong@ujs.edu.cn

5.1 Introduction to Microbial Electrochemical Systems

Electrochemical systems, although one of the oldest fields of chemistry, remains one of the most active research fields that continuously provides tremendous and facile solutions to fulfill the world's demands for energy[1-4] (*e.g.* chemical-/bio-fuel cells) and pollution treatment,[5] which are vital to the sustainability of the earth. Chemical fuel cells are one of the most successful applications of electrochemistry. However, traditional chemical fuel cells are often operated at high temperatures using expensive and rare metal catalysts for oxidation. Moreover, their need for very specific fuels with high purity and high cost largely limits their practical applications. Electrochemical wastewater treatment is another application of electrochemical systems, but it has inherent drawbacks, such as intensive energy consumption, high cost,

RSC Green Chemistry No. 34
Chemical Biotechnology and Bioengineering
By Xuhong Qian, Zhenjiang Zhao, Yufang Xu, Jianhe Xu, Y.-H. Percival Zhang, Jingyan Zhang, Yangchun Yong, and Fengxian Hu
© X.-H. Qian, Z.-J. Zhao, Y.-F. Xu, J.-H. Xu, Y.-H. P. Zhang, J.-Y. Zhang, Y.-C. Yong and F.-X. Hu, 2015
Published by the Royal Society of Chemistry, www.rsc.org

and low efficiency.[6] In contrast, microbial catalysts (microorganisms) provide an alternative as they have greater flexibility in that they can be operated at mild temperatures and pressures using diverse organic fuels without rare and expensive metal catalysts.[6,7]

Thus, microbial electrochemical systems (MES), which are defined as electrochemical systems that use whole-cell microorganisms (*e.g.* bacteria, fungi) as the catalysts, have been developed and have become a relatively new and hot research field.[6–8] By taking advantage of the great diversity of microbial metabolism, a wide variety of fascinating processes become practically possible.[9] Thus, different kinds of MES, originally inspired by the interesting physiological phenomenon of electron exchange between cells and electrodes, have been demonstrated, *e.g.*, microbial fuel cells (MFC), microbial electrosynthesis cells (MESy), and microbial desalination cells (MDC) (Figure 5.1).[9] The main purposes of MES research are to provide sustainable and energy-saving solutions for wastewater treatment, CO_2 biotransformation, bioenergy production, *etc.*[10–14] Actually, MES has been one of the most booming research fields in environmental science, biotechnology, material science, and related fields.[9,11,15,16] During the past few decades, the performance of MES has improved drastically through the efforts of multidisciplinary research, in which chemical biotechnology approaches have contributed greatly. In this chapter, the chemistry and biology of MES,

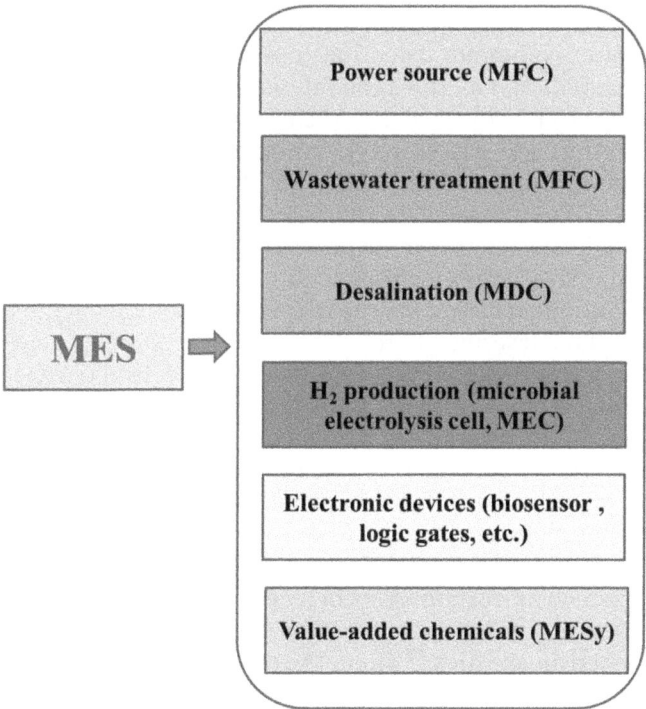

Figure 5.1 Potential applications of MES technology.

as well as the role of chemical biotechnology in improving extracellular electron transfer (EET) and whole-cell bioelectrocatalysis, are thoroughly reviewed. The future directions for chemical biotechnology in MES are also discussed.

5.2 Chemistry of MES

Basically, MES are composed of a series of chemical reactions/processes catalyzed by microorganisms. These processes include substrate decomposition/metabolism, electron releasing, electron transfer, energy production, and synthesis of electrocommodities. In this section, the chemistry of MES systems will be briefly described by taking two typical MES systems (*i.e.*, MFC and MESy) as examples.

5.2.1 Chemistry of MFC for Energy Harvesting from Waste

Microbial fuel cells (MFC), which were invented over 100 years ago,[17,18] are the most extensively studied MES and have attracted widespread interest during the last few decades.[16,19,20] By using MFC, the chemical potential energy stored in an organic substrate can be released by microbial metabolism and converted to electrical energy. A MFC device usually consists of two major components: the anode and the cathode. This process often contains anodic oxidation (substrate oxidation and electron transfer) and cathodic reduction (Figure 5.2), *i.e.*, the substrate is oxidized by the microorganism and separated into negatively charged electrons and positively charged protons; the electrons are passed to the anodic electrode and then delivered to the cathode through an electrical wire, thereby generating an electrical current; the separated protons in the anodic chamber are transferred to the cathode through the electrolyte or proton exchange membrane (PEM), and eventually reunite with the electrons at the cathode.[6,19] In all, the anode reaction and the cathode reaction are the main and essential chemical processes of MFC and will be described here.

The anodic reaction of a MFC is the process of electron release from a substrate, called substrate oxidation by microorganisms (Figures 5.2 and 5.3). In a MFC system, the anodic chamber is operated in anaerobic conditions, where the substrates are metabolized by the microorganisms with anaerobic metabolic pathways.[21,22] It was proposed that the substrate oxidation catalyzed by metabolic enzymes is accompanied by NAD^+/FAD^+ reduction to $NADH/FADH_2$. The $NADH/FADH_2$ is considered to be the intracellular electron carrier, which is responsible for accepting the electrons from cellular metabolism and passing them to the inner membrane redox proteins/enzymes or electron shuttles.[22] Then, the electrons are transferred *via* extracellular electron transfer pathways and finally passed to the electrode (this will be described in detail in the following section) (Figure 5.3).[23,24] The

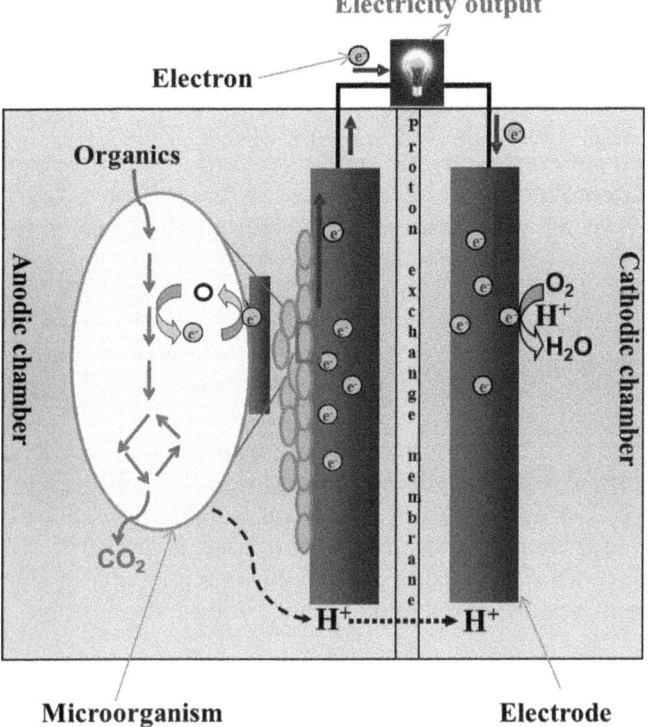

Figure 5.2   Simple model of a MFC.

typical anodic reactions in MFC systems $(nCH_2O + nH_2O \rightarrow nCO_2 + 4ne + 4nH^+)$ are summarized as follows:

Acetate $\rightarrow CO_2 + H^+ + e$ (for *Geobacter*-inoculated MFC fueled by acetate)[25]

Lactate $\rightarrow$ acetate $+ H^+ + e$ (for *Shewanella*-inoculated MFC fueled by lactate)[26]

Glucose $\rightarrow$ lactate $+$ succinate $+$ formate $+ H^+ + e$ (for *E. coli*-inoculated MFC fueled by glucose)[22]

The cathodic reaction of a MFC is an electron consumption or reduction process (Figure 5.2).[27] For a typical air cathode MFC, Pt is usually used as the catalyst for oxygen reduction.[27,28] Under Pt catalysis, the electrons from the anode and protons transported from the anodic compartment will reduce the oxygen, which is accompanied by water generation.[27] The chemical reaction occurring on the air cathode is $O_2 + H^+ + e \rightarrow H_2O$ (Figure 5.2). Recently, biocathodes besides metal or chemical cathodes have been developed by employing microorganisms to catalyze the cathode reaction.[29] By using bio-cathodes, other substrates besides oxygen, such as CO_2 and nitrate, can be efficiently used for cathodic reduction.[30-32]

Based on the chemistry of MFC, a simultaneous pollutant treatment (organic waste decomposition in the anode compartment, or nitrate/sulfate reduction in the cathode compartment) and energy generation (electric current) process has been demonstrated and has become the main focus of MFC research.[33] During the past few decades, various wastewaters (domestic or industrial) have been used as MFC substrates for electricity generation and waste treatment. However, low power output is still the main concern for practical application. Therefore, great efforts have been made to improve the power output of MFCs through genetic engineering,[22,28,34] microbial community acclimation,[35-37] electrode decoration or new electrode design,[38-40] and MFC architecture design.[41,42] The power output of MFCs has increased about 10 fold during the past decade. To date, the maximum power density obtained for MFCs is ~2 to 3 W m^{-2},[9] while the maximum volume power output produced is up to 1.55 kW m^{-3}.[43] However, it still cannot meet the requirements for practical application. Even worse, continued performance improvement is becoming more and more difficult by using traditional strategies such as electrode modification, strain acclimation, and chamber configuration design. The lack of

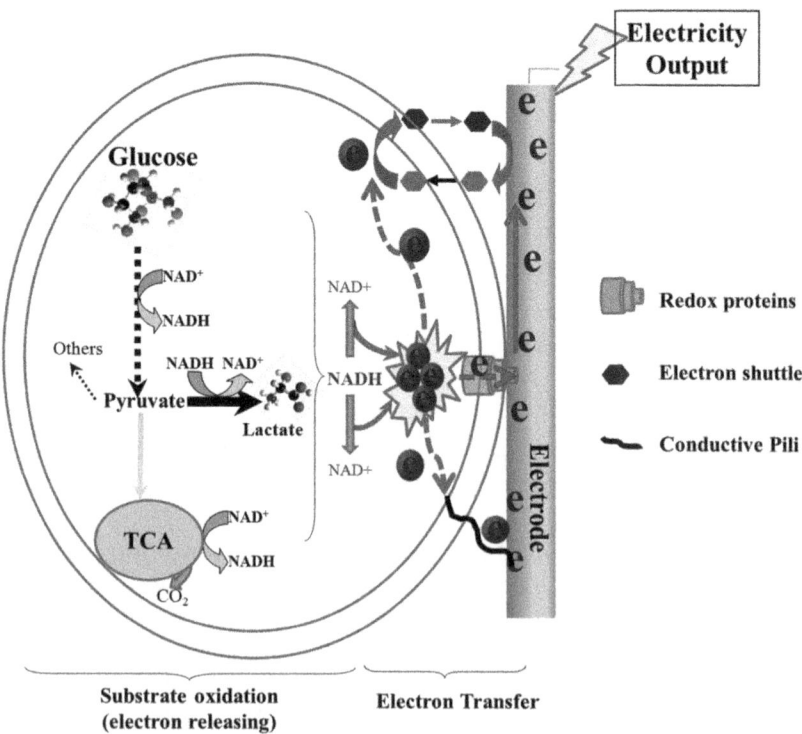

Figure 5.3 Proposed mechanisms for electron release and electron transfer in a MES system.

knowledge on the detailed biological mechanisms of MFCs is the main obstacle hindering the improvement of this technology. So, efforts at exploring the unclear biological mechanisms and the development of new manipulating strategies should be emphasized. Investigation of the biological mechanisms is of special importance as the cellular behaviours play key roles in this MFC technology. For example, microbial metabolism is the initial driving force for electron releasing and the electric interaction between the bacteria and the electrode is essential for the success of electron extraction by the anode. More importantly, exploration of the mechanisms would provide us with new insights and thus the possibility to develop new strategies to improve MFC performance, which is expected to lead to new breakthroughs. In addition, more attention should be paid to other low-power consuming applications, such as biosensors, portable power sources, and other bioelectronics, which may accelerate the practical applications of MFC.

5.2.2 Chemistry of MESy for Electrocommodities Production

In light of the fast development and deeper understanding of MFC technology, other related new technologies/processes have been proposed and demonstrated in recent years.[17] MES technology has evolved from MFCs devoted to bioenergy harvesting to various new microbial electrochemical devices with a multitude of specialized applications.[9] Microbial electrosynthesis (MESy) refers to microbial synthesis of fuels and chemicals driven by electricity and has attracted much attention in the biotechnology field in recent years.[9,11] MESy was established based on the discovery of microbial electron uptake and electricity-driven reduction of CO_2 by a cathodic microbial community.[44] The up-taken electrons provide reducing power and influence the cell metabolism, biochemical yield, cellular ATP yield, and cellular growth rate of the microorganisms.[45] More impressively, some acetogenic bacteria, such as *Sporomusa ovata*, attached on the cathode can produce acetate or butyrate by using CO_2 and electricity as the sole carbon and electron donors, respectively.[44] Theoretically, a MESy system can produce a variety of value-added organic compounds from CO_2 or other organic substrates.[7,46] Figure 5.4 lists

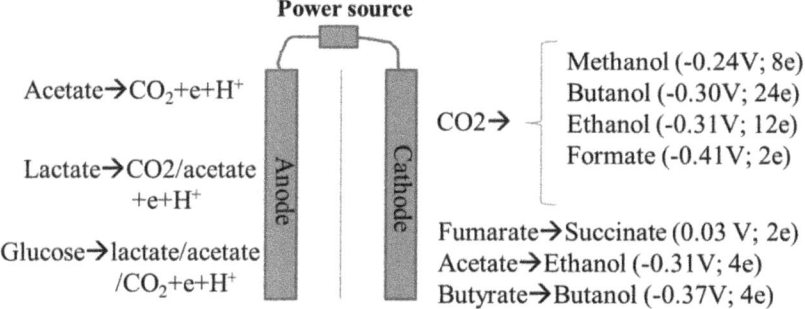

Figure 5.4 Potential organics that can be produced on the cathode of a MESy system.

the potential organics that can be produced on the cathode of a MESy system. There are three main fundamental processes involved for MESy:

1. Electron up-take;
2. Converting the up-taken electrons to intracellular reducing equivalents; and
3. CO_2 biotransformation/chemical synthesis.

Experimental evidence implies that the electrons from the cathodic electrode are transported into cells with different unclear electron transfer routes, and then elevate the intracellular reducing equivalents (*i.e.*, $NADH^+/NAD^+$), which forces the CO_2 reduction to more complicated organics through different metabolic pathways. Although several inwards electron transfer pathways and CO_2 biotransformation pathways have been proposed,[45,47,48] the detailed mechanisms are still ambiguous and more efforts should be made at understanding them.[49]

By using CO_2 as the sole carbon source, several low molecular organics (such as acetate, methane, butyrate, and formate) have been produced with the MESy process.[44,50,51] As CO_2 is an extensively oxidized compound and is ubiquitous, it is a good electron acceptor for MESy. In a typical MESy process, the up-taken electrons will elevate the ratio of $NADH^+/NAD^+$, and facilitate the reduction of CO_2 with various microbial metabolic pathways.[9,11] The advantages of MESy starting with CO_2 are the possibility to fix excessive greenhouse gas (CO_2) and store electricity as chemicals or fuels.[11] However, the main limitation of this strategy is low production rates and titers as well as a high ratio of by-products. Thus, improvement of production rates and reduction of by-products are essential for the practical application of MESy. Efforts at cathode modification[39,52–54] and long-term consortia acclimation[55] have been made, and these have efficiently enhanced the production rates and titers. Strain improvement or process optimization are also expected to be promising strategies to improve the MESy process, and deserve more attention. The disadvantage of CO_2 as the starting building block is the large electron requirement and the involvement of multiple steps (usually resulting in large efficiency losses) for the synthesis of compounds with long carbon chains (chemicals or fuels).[11] Thus, energy efficiency, target compound selection, synthesis pathway optimization/design, and process optimization are vital and should be taken into account to develop a cost-effective and sustainable MESy process starting with CO_2.

By using waste organics as the starting building blocks, biofuel chemicals, such as ethanol,[56] butanol,[57] and succinate,[39] have been produced *via* MESy. For example, isobutanol and 3-methyl-1-butanol were successively produced with a rationally designed un-native metabolic pathway using electrochemically reduced formic acid as the starting carbon source and electron carrier.[57] Another typical MESy process starting with waste organics is the cathodic reduction of acetate to ethanol. By using a mixed microbial community attached on the cathode with a poised potential of −550 mV (*vs.* NHE), ethanol production from acetate was reached a concentration of 1.82 mM with high production efficiency (~75%) and coulombic efficiency (~49%).[56] The advantages of this process are the upgrading of low-concentration raw organic waste (which is ubiquitous in wastewater) and the storage of electricity.[9,11] Moreover, this process is also expected to reach higher energy efficiency and production rates over the CO_2 process due to the lower electron requirements and reduced reaction steps.

5.2.3 Chemistry of Extracellular Electron Transfer

As the cell membrane of microorganisms is non-conductive, how to transfer the electrons across the cell membrane (usually termed as extracellular electron transfer (EET)) is the key issue for MES.[9,16] For the bioanode, extracellular electron transfer usually refers to outwards electron transfer, which is related to transportation of intracellular electrons to the solid electrode. To date, at least three electron transfer pathways have been explored, *i.e.*, electron shuttle mediated electron transfer, outer membrane redox proteins mediated contact-based electron transfer, and conductive pili mediated electron transfer (Figure 5.5).[15,16]

It was found that some small redox-active chemicals (termed electron shuttles) can diffuse across the cell membrane and serve as electron carriers to assist the electron transfer from the bacteria to the electrode. The electron shuttle mediated electron transfer process usually contains three steps, *i.e.*, being reduced by cells (the electron shuttle is converted to a reductive state),

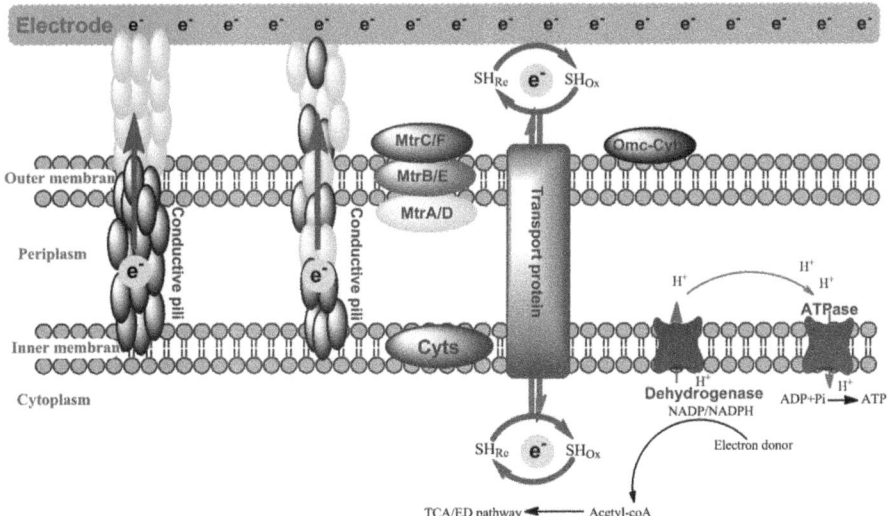

Figure 5.5 Simple model for extracellular electron transfer between cells and an electrode. SH_{Re}: electron shuttles in redox state. SH_{Ox}: electron shuttles in oxide state.

transportation (move across the cell membrane and diffuse to the electrode surface), and reducing the electrode (the electron shuttle is converted to an oxidative state and the electrons are passed to the electrode). In detail, the electron shuttle in the oxidative state diffuses into cells or interacts with the membrane redox protein and is reduced, then the reduced electron shuttle diffuses outside the cells and interacts with the electrode; the electrons stored in the electron shuttle are then passed to the electrode and meanwhile the electron shuttle is oxidized to its oxidative state, which is ready for re-reduction by cells. To date, many bacteria have been found to be able to produce electron shuttles that directly engage in electron exchange with the electrode, *e.g.*, flavin produced by *Shewanella*[58] or phenazines produced by *Pseudomonas*.[34,59] In addition, synthetic molecules, such as neutral red,[60] naphthoquinone,[61] and methyl viologen, can also be used as electron shuttles.[62]

Although the whole-cell membrane is non-conductive, there are several redox proteins anchored on/in the membrane that confer nano-scale conductivity to the membrane and directly enable electron transfer across the cell membrane. These proteins usually assemble together in the periplasm and/or on/across the outer-surface membrane and act as an electron transfer chain to relay the electron across the membrane.[6,16] For example, the membrane-bound electron transfer chain of *Shewanella oneidensis* is a *trans* icosa-heme complex, MtrCAB, that can move electrons across the membrane.[63] The MtrC is a decaheme cytochrome located on the outside of the outer cell membrane that mediates the electron transfer to the extracellular substrate (*e.g.*, solid electrode). MtrAB is the transmembrane electron transfer module that is responsible for electron transport from the periplasm to MtrC.[11,63,64] More interestingly, recent findings indicates that this electron conduit is capable of reverse electron transfer, *i.e.*, electron up-take from extracellular electrodes.[47,65]

Another interesting electron transfer route is conductive pili (also called nanowire) bridged electron exchange between the bacteria and the electrode. By using conducting-probe atomic force microscopy or a nanoelectrode platform, it was surprisingly found that the pili produced by several bacterial species are highly conductive (1–5 mS cm^{-1}, which is comparable to synthetic metallic nanostructures).[66–69] Interestingly, these conductive nanowires play crucial roles in electron transfer between the cells and the electrode. For example, the pili-deficient mutant of *Geobacter sulfurreducens* cannot transport intracellular electrons to metal oxides, although their cell attachment is not significantly affected.[67] More impressively, the nanowires can conduct over centimeter scale distances (~1000 times of the size of a bacterium). This unique property is vital to the long-range conductivity of biofilm matrices.[69] Very recently, charge propagation along individual pili proteins was visualized by using ambient electrostatic force microscopy and a delocalized charge transfer mechanism was proposed.[70] However, the detailed mechanism for electron transfer along the nanowires still needs more investigation.

One exoelectrogenic microorganism usually has more than one electron transfer route to exchange electrons between the cells and the electrode. For example, *S. oneidensis* MR-1 contains all three pathways for electron transfer.[39] It can synthesize flavins and secret into the medium, which serve as electron shuttles.[58] *S. oneidensis* MR-1 also has electroactive outer membrane proteins called c-type cytochromes (CymA, MtrCAB, *etc.*) which are composed of an electron transport chain across the cell membrane.[63] Moreover, this bacterium also produces conductive pili, which are also involved in electron transfer.[68] By using the nanowires, bacteria may be able to form a conductive biofilm matrix and directly transport the intracellular electrons from the outer surface of the biofilm.

5.3 Chemical Bioengineering of Extracellular Electron Transfer in MES

5.3.1 Chemical Electron Shuttles Activate Electron Transfer

High electron discharge capabilities of electrochemically active microbes have a significant impact on the performance of the corresponding MES. To date, three kinds of mechanisms for extracellular electron transfer (EET) involved in MES have been established:

1. Direct electron transfer *via* c-type cytochromes proteins on the outer membrane of the bacteria;
2. Direct electron transfer *via* conductive pili; and
3. Indirect electron transfer based on the electron shuttle mediated process.

Based on these mechanisms (shown in Figures 5.3 and 5.5), several chemical engineering strategies to accelerate EET are put forward.[24,71]

The basic process of EET is a bottleneck that governs the power output of MFCs. One of the most general electron transfer pathways is electron shuttle-mediated EET. Many electrochemically active microbes, such as *Shewanella oneidensis*, *Geobacter sulfurreducens*, *Pseudomonas aeruginosa*, and *Escherichia coli*, can endogenously produce redox chemical molecules that are known as electron shuttles (Table 5.1). The electron shuttles can transfer the electrons from bacterial metabolism towards the electrode over a much longer distance than the electron traverse distance. Recent studies have been able to detect redox-active compounds, such as flavins secreted by microbes to mediate electron flow to metals, located >50 μm away from cell surfaces. Marsili *et al.* utilized a poised electrode as an electron acceptor and detected redox-active molecules within biofilms. The process of flavin secretion, shuttling, and binding demonstrates that biofilms of *Shewanella* use generated flavins in mediating electron transfer to external acceptors. The detected concentration gradient of redox-active molecules can even guide cells to favorable sites.[58,72] Other studies show that endogenous flavin molecules enhance the ability of bacterial outer-membrane c-type cytochromes (Omc-Cyts) to transport electrons as redox cofactors in the form of bound flavin semiquinones, but not free flavins.[163]

In addition to self-generated shuttles, synthetic compounds with similar chemical structures also show the same function. Changing the molecular structure of phenazine-type redox mediators by artificial synthesis can significantly influence microbial extracellular electron transfer.[73] Immobilization

Table 5.1   Electron shuttles used in MES systems.

Exoelectrogenic strains	Chemical electron shuttles (self-generated)	Chemical electron shuttles (exogenously added)	Ref.
Pseudomonas aeruginosa KRP1	Phenazines		59, 78 and 83
Shewanella oneidensis MR-1	Flavins (riboflavin, flavin mononucleotide, or flavin adenine dinucleotide)		84
Shewanella oneidensis MR-1	Flavin semiquinones, c-type cytochrome CymA		85 and 86
Shewanella sp. MR-4	Flavins		58
Geobacter sulfurreducens		Cysteine, anthraquinone-2, 6-disulfonate (AQDS)	87
Geobacter sulfurreducens JS-1	Extracellular cytochrome		81
Geobacter species	Flavins		88
Saccharomyces cerevisiae PTCC 5269		Neutral red, thionine, ferric chelate, and methylene blue	89
Escherichia coli		Riboflavin and humic acid	90
Faecalibacterium prausnitzii	Flavins and thiols		91
Lactococcus lactis		Flavins	92

of a small active phenothiazine derivative, methylene blue, on the electrode surface can accelerate the microbial EET due to the enhanced interactions between the bacterial outer-membrane cytochromes and the immobilized methylene blue.[74]

Electron shuttles serve as the intermediate molecule for electron transport between the electrode and the redox enzyme in the microorganism. The key features of these chemical electron shuttles have been well established. First of all, the molecules should be able to form redox couples. This character is indispensable as it enables them to store and release electrons, which are vital for electron shuttling. Shuttles in the oxidized state can be reduced in the presence of electrons within the cell membrane and then pass through the membrane and release the electrons to the anode and return to the oxidized state simultaneously. This redox cycle improves the EET and thus enhances the power output. Secondly, both of their reduced and oxidized forms should be stable and non-degradable. *Shewanella* species have been proved to be able to accumulate flavins in the supernatants. External flavins at biologically relevant concentrations primarily act as electron shuttles for *S. oneidensis* MR-1.[75] Finally, the shuttles should be nontoxic to the microbes.

Several strategies to increase the production of electron shuttles have been developed to improve the MFC performance in the model exoelectrogens. For *Shewanella* species, flavins (riboflavin and flavin mononucleotide) are the most well-known self-secreted electron shuttles. Using deletion mutants lacking various Mtr-associated proteins, the significance of the Mtr extracellular respiratory pathway for the reduction of flavins has been demonstrated. The decaheme cytochromes found on the outer surface of the cell (MtrC and OmcA) are required for the majority of Mtr-associated proteins' activity.[75] Weakly acidic pH resulted in poor performance of the MFC and low riboflavin concentrations in the bacterial cultures,[76] while enhanced electrochemical activity of riboflavin was reported at alkaline pH. The increase of riboflavin biosynthesis by *Shewanella* at the alkaline condition underlies the improvement in the electricity output in MFCs.[77]

For *Pseudomonas aeruginosa*, another famous exoelectrogenic Gram-negative bacteria, phenazines are the main naturally secreted shuttles. The biosynthesis of the redox shuttles is regulated by the 2-heptyl-3, 4-dihydroxyquinoline (PQS) quorum-sensing (QS) system, which inhibits the anaerobic growth of *P. aeruginosa*. When over-expressing the PqsE effector in a PQS-negative *ΔpqsC* mutant, the engineered strain exhibited improved MFC electrical performance and correlated with an over-production of phenazines.[78] The manipulation of electron shuttle synthesis pathways could be an efficient approach to improve the power output of MFCs. By overexpression of *phzM* (methyltransferase encoding gene, the key biosynthesis gene for pyocyanin), the maximum power density of *P. aeruginosa-phzM*-inoculated MFC was enhanced four fold compared to the original strain. In addition, the strain exhibited a 1.6 fold increase in pyocyanin (PYO) production.[79] Pyocyanin, a well-studied phenazine shuttle, can also be utilized to enhance EET and growth by other bacteria.[59] With the genetic overexpression of the *rhl* QS system, the phenazine electron shuttles were increased, which in turn significantly improved the electrocatalytic activity and the electricity output of the MFC.[34]

Geobacter sulfurreducens is also one of the most extensively studied electrogenic microorganisms. It is representative of the *Geobacter* species, which is a class of electrochemically active microbes commonly found in environmental samples. Its genome sequence is available and is amenable to genetic manipulation.[72,80-82] Previous reports considered that this bacterium relied on conductive pili for EET. Very recently, it was surprisingly found that deletion of the gene encoding PilA, the structural pilin protein, did not significantly affect its ability to reduce Fe(III) oxide. Further analysis showed that *G. sulfurreducens* released a distinct protein electron shuttle—PgcA—which is involved in EET and Fe(III) oxide reduction.[164] The electron shuttles accompanied with biogenic iron species also contributed to the transformation of carbon tetrachloride in *G. sulfurreducens*.[165]

Overall, these model exoelectrogens generally have clear physiological, biochemical, and genetic backgrounds. They provide nice research platforms and act as critical tools for exploring EET mechanisms.

5.3.2   Conductive Polymer-Bridged Direct Electron Transfer

One efficient engineering strategy to improve the EET is to use the bridging action of conductive polymers. Such polymers can deliver electrons directly from cell to cell/electrode, which is similar to the natural direct electron transfer (DET) between cells and the electrode.

DET depends on membrane bound redox-active proteins or conductive pili (nanowires). It is the direct physical contact of the microbe with the anode and no redox species or electronic shuttles are involved. A wide diversity of exoelectrogens shows effective DET mechanism with electrochemical and genetic evidence. Bacterial nanowires act as a viable microbial strategy for EET and are produced on the exoelectrogenic cell surface. Bacterial nanowires are conductive pilus-like appendages in direct response to electron-acceptor limitation. The Msh pilin biosynthesis system was identified as an integral pathway for DET in *S. oneidensis* MR-1.[96] Among microbial nanowires in various electrochemical active microorganisms, the type IV pili of *G. sulfurreducens* and *G. metallireducens* were the only filaments that had been shown to be required for DET to extracellular electron acceptors or for conduction of electrons through biofilms.[97] Another form of DET is probable through membrane bound cytochromes. Direct electrochemistry of OmcA, an outer-membrane c-type cytochrome, on Fe_2O_3 electrodes was observed using cyclic voltammetry.[93] Many other studies indicate that c-type cytochromes play a predominant role in the DET.[94,95] The possible roles of outer membrane c-type cytochrome (*e.g.* OmcZ, OmcB, OmcE, and OmcS), type IV pili, and protons in total EET have been revealed in the wild type and mutant *G. sulfurreducens*: OmcZ and OmcS take part in helping the shuttles to pass through the biofilm bulk; OmcB mediates the shuttles across the biofilm/electrode interface; type IV pili are important in both the above processes while OmcE is involved in neither reaction; protons produced from acetate oxidation act as charge compensating ions in the process of the shuttles passing through the biofilm bulk; diffusion of protons within the biofilm would decelerate this process.[93–97]

Therefore, it can be concluded that the redox proteins (c-type cytochrome *etc.*) and electrodes comprise the main part of DET. The increased contact of the redox proteins (c-type cytochrome *etc.*) and electrodes would be helpful to DET, thereby influencing the MFC efficiency.[98] Evidence has revealed that some conductive polymers could act as electrical wiring between cells and anodes (Table 5.2).

Poly(1-vinylimidazole)$_{12}$-[Os-(4,4'-dimethyl-2,2'-bipyridyl)$_2$Cl$_2$]$^{+/2+}$ and poly(vinylpyridine)-[Os-(*N*,*N*'-methylated-2,2'-biimidazole)$_3$]$^{2+/3+}$ were reported for their efficient capability of mediating electrons transfer between bacterial cells to electrodes.[99,100] With *S. oneidensis*, the osmium redox polymer modified anode showed a 4-fold increase in current generation and a significant decrease in the start-up time for electrocatalysis.[101,102] Using an anode modified with electropolymerized polypyrrole, a dramatic improvement in energy output was noticed in the MFCs.[103] MFCs with a polypyrrole/anthraquinone-2,6-disulfonic disodium salt (PPy/AQDS)-modified anode

Table 5.2 Related studies of anode modification with conductive polymers in MFCs (P.D.: power density. C.D.: current density).

Original Anode	Anode after modification	Cathode	Biocatalyst	Type of MFC	Fuel	MFC efficiency after anode modification	MFC efficiency before anode modification	Ref.
Carbon paste electrode	Immobilized cell on carbon nanotube modified carbon paste electrodes by poly(1-vinylimidazole)$_{12}$-[Os-(4,4'-dimethyl-2,2'-dipyridyl)$_2$Cl$_2$]$^{2+/+}$	Pt wire	*Pseudomonas putida* DSM 50026	Single-compartment MFC reactors	Glucose	C.D. ≈ 280 nA cm^{-2} (response to 2 mM glucose, with CNT and Os)	C.D. ≈ 130 nA cm^{-2}; (response to 2 mM glucose, without CNT and with Os)	109
Carbon felt electrode	Polypyrrole/anthraquinone-2,6-disulphonic disodium salt (PPy/AQDS)-modified carbon felt anode	A bare carbon felt electrode	*Shewanella decolorationis* S12	Dual-chamber MFC	Lactic Acid	P.D. = 1300 mW m^{-2}; C.D. = 2.74 A m^{-2}	P.D. = 100 mW m^{-2}; C.D. = 0.5 A m^{-2}	110
Carbon paper (TP) electrode	Multi-wall carbon nanotube (MWNT) and polyelectrolyte polyethyleneimine (PEI) were employed to modify a carbon paper (TP) electrode utilizing a layer-by-layer (LBL) assembly technique	PTFE-treated wet-proofed carbon paper coated with Pt (0.5 mg cm^{-2}) on one side was used as an air cathode	Mixed-culture anaerobic granular sludge	Single-compartment MFC	Glucose	P.D. = 290 mW m^{-2}, C.D. = 568 mA m^{-2}; CV = 511 mV	P.D. = 241 mW m^{-2}, C.D. = 448 mA m^{-2}; CV = 538 mV	111
Carbon cloth electrode	Polyaniline hybridized three-dimensional (3D) graphene	Carbon cloth	*Shewanella oneidensis* MR-1	Two-chamber MFC	Lactate	P.D. = 768 mW m^{-2}	P.D. = 158 mW m^{-2}	38

(continued)

Table 5.2 (*continued*)

Original Anode	Anode after modification	Cathode	Biocatalyst	Type of MFC	Fuel	MFC efficiency after anode modification	MFC efficiency before anode modification	Ref.
Traditional 3D anodes such as reticulated vitreous carbon (RVC), graphite plate (GP), carbon felt (CF), and graphene-coated sponge (GS) electrodes	Polyaniline-hybridized loofah sponges	30% wet-proofed carbon cloth with four layers of PTFE coating	Activated anaerobic sludge	Single-chamber MFC	Sodium acetate	P.D. = 1098 mW m^{-2}	P.D. for CF = 572 mW m^{-2}; P.D. for GP = 383 mW m^{-2}; P.D. for RVC = 650 mW m^{-2}; P.D. for GS = 612 mW m^{-2}; P.D. for LSC = 701 mW m^{-2}	112
Graphite electrode	Electrospun carbon nanofibers obtained from polyacrylonitrile (PAN) and PAN blends with either activated carbon (PAN–AC) or graphite (PAN–GR)	Bare graphite	*Shewanella oneidensis* MR-1	Single-chamber MFC	Sodium lactate	C.D. for PAN = 115 µA cm^{-2}; C.D. for PAN–AC = 139 µA cm^{-2}; C.D. for PAN–GR = 155 µA cm^{-2}	C.D. = 11.5 µA cm^{-2}	113
Graphite felt electrode	One-step electrosynthesis of polypyrrole (PPy)/graphene oxide (GO) composites on the graphite felt	Pt mesh	*Shewanella oneidensis* MR-1	Dual-chamber MFC	Lactate	P.D. = 1326 mW m^{-2}; C.D. = 7.53 A m^{-2}	P.D. = 166 mW m^{-2}; C.D. = 2.35 A m^{-2}	114
Carbon paper electrode	Polyaniline modified by ethylenediamine	Carbon paper coated with 0.5 mg cm^{-2} Pt	Palm oil mill effluent (POME) anaerobic sludge	Two-chamber MFC	Glucose	P.D. = 136.2 mW m^{-2}; C.D. = 3.37 A m^{-2}	P.D. = 92.4 mW m^{-2}; C.D. = 2.77 A m^{-2}	106

exhibited higher performance. Electrochemical analysis results suggested that the increased surface area due to anode modification, the enhanced electron-transfer efficiency from the bacteria to the anode, and the increased number of bacteria attached to the anode explained the high power generation.[166] Evidence indicates that the microstructure and the conductivity of these polymers is amenable for bridging from the cell to the electrode. A conductive artificial biofilm was designed as a conductive matrix interlocked with polypyrrole, within which *S. oneidensis* was wrapped with micro-sized graphite. The PPy polymeric chains and graphite particles provided a convenient way for the electrons produced by the microbe's metabolites to transfer to the anode by EET. The MFCs with this artificial biofilm showed an 11-fold increase in energy generation compared with natural biofilms.[26]

Another widely studied conducting polymer used in MFCs is polyaniline (PANI). Its hydrophilicity endows it with increased biocompatibility with anode materials. Further research has reported that doped polyaniline modification of the anode could be an effective strategy to improve the energy output of MFCs.[46,104-106] When polyaniline nanowire arrays (PANI-NA) were used as the working electrode in a three-electrode system, multiple oxidation levels of PANI at different applied potentials lead to multiple abilities in mediating EET. The aligned nanostructures enable enhanced local topological interactions with microbes and hence efficient electrical interactions.[107] The biopolymer facilitates the EET from the bacteria to the anodes. Their formed scaffolds have a large specific surface area, which is also helpful for good bio-compatibility and nutrient transport for bacterial growth. When the MFCs were equipped with the PANI hybridized 3D graphene anode, a maximum power density of 768 mW m^{-2} was achieved, which was about four fold higher than that of carbon cloth anodes. Such three-dimensional macroporous structures provide multiplexed and conductive pathways and facilitate the EET.[38] Related studies of anode modification with conductive polymers in MFCs are summarized in Table 5.2.[108]

5.3.3 Carbon Nanomaterial-Promoted Direct Electron Transfer

According to the DET mechanism, the anode as the carrier of microbial adherence directly influences the contact between the bacteria and the anode and the overall performance of the MFC. It is considered to be the most important part of the system. A perfect anode has some basic requirements, such as good electronic conductivity, high specific surface area, good biocompatibility, inert, good stability, durability, low cost, and non-toxic. Nanostructured carbon materials, *e.g.* graphene, carbon nanotube (CNT), nanoribbons, and so on, generally have better conductivity and enhance the specific surface area of electrodes and are thereby widely used as electrode modifiers for improving the performance of MES. In addition, the roughness and dimensionality of nanomaterials is suitable for increasing the biofilm loading and enhancing the interaction between the bacteria and the electrode, and thus improving the EET. All of these characteristics have improved the performance of MFCs (Table 5.3).[115,116]

Table 5.3 Related studies of anode modification with carbon nanomaterials in MFCs (P.D.: power density. C.D.: current density).

Original anode	Modified anode	Cathode	Biocatalyst	Type of MFC	Fuel	MFC efficiency after anode modification	MFC efficiency before anode modification	Ref.
Carbon cloth	Bacteria-entrapped carbon nanoparticle paste on the carbon cloth	Carbon cloth	*Proteus vulgaris*	Two-chamber MFC	Glucose	P.D. = 269 mW m^{-2}	P.D. = 20 mW m^{-2}	117
Stainless steel mesh	Graphene powder was mixed with PTFE then coated on the surface of a stainless steel mesh	Carbon paper	*Escherichia coli*	Two-chamber MFC	Glucose	P.D. = 2668 mW m^{-2}	P.D. = 142 mW m^{-2}	46
Plain carbon paper	CNT powders	Plain carbon paper	*Geobacter sulfurreducens*	Two-chamber MFC	Sodium acetate	Anodic resistances = 173 Ω; output voltage = 650 mV		118
Carbon cloth	Carbon nanotube–textile (CNT-textile) composite	Carbon cloth	Domestic wastewater	Two-chamber MFC	Glucose	P.D. = 1098 mW m^{-2}; CD = 7.2 A m^{-2}	P.D. = 655 mW m^{-2}; C.D. = 2.8 A m^{-2}	119
Carbon cloth	Hierarchically porous chitosan/vacuum-stripped graphene scaffold	Carbon cloth	Domestic waste	Dual-chamber MFC	Glucose	P.D. = 1530 mW m^{-2}	P.D. = 19.6 mW m^{-2}	120
Graphite felt	Graphite sheet	Carbon paper	*Escherichia coli*	Two-chamber MFC	Glucose	P.D. = 2249 mW m^{-2}	P.D. = 100 mW m^{-2}	121 and 110

Carbon cloth	Electrochemically reducing graphene oxide first and coating polyaniline nanofibers afterwards on the surface of the carbon cloth	Carbon felt	Anaerobic sludge	Two cylindrical compartments MFC	Sodium acetate	P.D. = 1390 mW m^{-2}	P.D. = 463 mW m^{-2}	122
Carbon felt electrodes	Carbon felt-supported nano-molybdenum carbide (Mo$_2$C)/carbon nanotubes (CNTs) composite	Carbon paper cathode was pasted with 0.3 mg cm^{-2} commercial Pt-catalyst (20 wt% Pt/C) in a mixture of polyvinylidene fluorine (PVDF) by a weight ratio of 65:15 in 0.8 mL N-methyl-2-pyrrolidone (NMP)	Escherichia coli	Single-chamber MFC	Glucose	P.D. = 1260 mW m^{-2}; C.D. = 11.7 A m^{-2}	P.D. = 80 mW m^{-2}; C.D. = 0.63 A m^{-2}	123
Glassy carbon electrode	Nanocomposite of multi-walled carbon tubes and tin oxide (MWCNTs/SnO$_2$) coated on the glassy carbon electrode	Pt rod	Escherichia coli	Two-chamber MFC	Glucose	P.D. = 1421 mW m^{-2}	P.D. = 458.4 mW m^{-2}	124
Carbon paper	30 wt% Fe$_3$O$_4$/CNT modified carbon paper	20 wt% Pt-coated carbon cloth	Escherichia coli	Two-chamber MFC	Glucose	P.D. = 831 mW m^{-2}	Almost zero	125

CNTs are often used in electrode modification because of their good bio-compatibility and superior conductivity. CNT-modified carbon paste electrodes facilitate stable binding of bacteria to the electrode surface and exhibit efficient EET between the bacteria and the anode.[167] The improvement can be due to the high local density of electronic states in CNTs.[167] Utilizing a layer-by-layer assembly technique, the multi-walled carbon nanotubes (MWNT) and polyelectrolyte polyethyleneimine (PEI) were modified on the carbon paper electrode. A 20% enhancement in the power density compared to the plain carbon paper anode was observed.[111] Addition of 0.5% (w/v) CNTs in the cell-immobilized alginate beads led to moving intracellular nitrobenzene reduction to an extracellular reaction and a 74% improvement in the nitrobenzene reduction efficiency. CNTs facilitate the EET and even alter the electron flow route.[126] Employing the Fe_3O_4/CNT nanocomposite for anode modification in a mediator-less MFC, the MFC exhibited a remarkably higher maximum power output (830 mW m^{-2} *vs.* almost zero) than the unmodified carbon anode.[125] The direct electron transfer was possibly achieved when the carbon nanoparticles closely contacted with the redox centers that were incorporated in/on the outer cell membrane.

Graphene is a virtually two-dimensional material with excellent conductivity that is emerging as a powerful electrode modifier for MES. Using graphene-modified stainless steel mesh (GSM) as the anode, the MFCs delivered a maximum power density of 2668 mW m^{-2}, which was 18-fold higher than that of MFCs equipped with a plain stainless steel mesh anode.[168] The improvement could be ascribed to the high surface area and enhanced bacteria attachment on the anode.[46] A technique of ice segregation induced self-assembly was used for preparing a 3D hierarchically porous chitosan/vacuum-stripped graphene scaffold as a bioanode. The optimized bioanode exhibited a much higher maximum power density of 1530 mW m^{-2}, which is 78-fold that of a carbon cloth anode.[120] Another macroporous and monolithic anode based on polyaniline hybridized 3D graphene also outperforms the planar carbon electrode.[38]

However, all the electrode modification processes might depend on complicated chemical/electrochemical reactions that usually require high temperatures, high pressures, toxic chemicals, or extensive energy consumption. Self-assembling of bacteria with nanomaterials onto the electrode or forming a bio-conductive matrix was demonstrated to be another novel electrode modification strategy with an easy operation process that can overcome the aforementioned issues. Simultaneously adding CNT powders and *Geobacter sulfurreducens* into the anode chamber of a MFC, self-assembling helped them to form a composite biofilm on the anode. The startup time of the MFC was significantly shortened and the power output was improved as compared with the MFC without CNT powders.[118] The MWNT-based layer-by-layer self-assembled electrode was reported to be promising for electricity production by MFC.[111,127] By mixing *S. oneidensis* MR-1 and hydrophilic graphene oxide (GO) in the anode chamber, a 3D macroporous conductive biofilm self-assembled. Surprisingly, bioreduction of non-conductive GO to conductive rGO by *S. oneidensis* MR-1 was achieved during this assembly process. The self-assembled 3D macroporous electroactive biofilm facilitated highly efficient bidirectional electron transfer between cells and electrodes. This unique biofilm delivered a 25 times higher outward current and 74 times higher inward current higher than the natural biofilms.[39] The work related to the modification of anodes *via* carbon nanomaterials is listed in Table 5.3.[108] Generally, incorporating carbon nanomaterials into bioelectrocatalysis systems is a promising approach to improve EET and the performance of MES.

5.4 Chemical Manipulation of Cell Physiology in MES

Cell physiology is a key factor that directly affects the cell gene expression, enzyme activity, cell metabolisms, cell growth, and viability. Manipulation of cell physiology through chemical strategies is considered to be an efficient and cost-effective way to improve the performance of various bioprocesses, such as secondary metabolites fermentation and pollutant biodegradation.[128–130] Therefore, attempts have been made to improve MES performance through the chemical manipulation of cell physiology.

5.4.1 Chemically Improved Cell Permeability for Efficient MES

As mentioned above, the electron transfer efficiency is largely affected by the permeability of the cell membrane.[131,132] Besides genetic engineering,[132] chemical treatment is an alternative and cost-effective way to improve membrane permeability. Surfactants are amphiphilic, surface active, organic compounds that lower the surface tension of the surface interface. It was found that surfactants could change the ultrastructure of cell membranes and increase membrane permeability. Surfactant addition is extensively used as an effective strategy to increase the membrane permeability to organic chemicals and thus enhance the degradation of pollutants or the secretion of products by microorganisms.[133–137] Recently, surfactants have also been used in MES systems for EET and electrocatalysis improvement. According to these studies, surfactant addition largely increases the COD removal, coulombic efficiency, and power/electricity output of MES systems by reducing electron transfer resistance across the cell membrane and enhancing electron shuttle production.[131,138–140]

Surfactants reduce electron transfer resistance across cell membranes. On addition of surfactants, the membrane permeability of electroactive bacteria was significantly increased, which in turn accelerated the electron shuttle transport across the cell membrane.[131] Moreover, the surfactant changed the cell roughness and increased the cell adhesion on the solid electrode, which resulted in higher biofilm loading.[140] For example, chemically synthesized surfactants ethylenediamine tetraacetic acid (EDTA), polyethyleneimine

(PEI), and Tween-80 were applied for improvement of microbial electrocatalysis in MFC.[138,140] *Pseudomonas aeruginosa* was treated with EDTA or PEI for 10 min before MFC inoculation, which resulted in cell shape changes, increased surface roughness, and the formation of nanoscale membrane pores. Moreover, the cell membrane permeability was significantly enhanced, as expected.[140] More interestingly, EDTA or PEI treatment improved the cell adhesion on the electrode. Thus, the chemically treated cells showed much higher electroactivity and delivered much higher power output as compared with the untreated cells.[140] In addition, both the overall MFC resistance and the anode resistance were significantly reduced by the addition of surfactants. In a MFC with mixed bacteria culture, on addition of Tween-80, the maximum power output, COD removal, and coulombic efficiency increased from 21.5 W m^{-3} to 187 W m^{-3}, from 57% to 73%, and 10.8% to 36.3%, respectively.[138] Moreover, the biofilm loading was significantly increased while the whole cell resistance and anode resistance were reduced.[138] However, its physiological effects on bacteria or bacterial communities are still unclear and need to be further investigated.

Chemically synthesized surfactants are usually toxic to bacteria, which largely limits their applications.[141] Biosurfactants produced by microorganisms with lower toxicity are alternatives to chemical surfactants for biological applications. Rhamnolipid, a typical biosurfactant mainly produced by *Pseudomonas*, is widely used in various fields.[142,143] With the addition of rhamnolipid, the internal resistance of MFC decreased from 29 Ω to 5 Ω, the open circuit voltage (OCV) increased from 483 mV to 878 mV, while the power output increased from 22 W m^{-3} to 275 W m^{-3}.[139] More recently, our group demonstrated that a new type of biosurfactant, sophorolipid (SOP, a glycolipid biosurfactant produced by non-pathogenic yeast), was able to improve electron transfer and MES efficiency through increasing cell membrane permeability and bacterial electric activity.[131] Under optimized conditions, SOP addition significantly increased the cell permeability as monitored by a fluorescent agent assay. In a single-chamber air-cathode MFC, the maximum power output delivered from the SOP treatment was about 4 times higher than that of the control (untreated).[131]

Surfactants enhance electron shuttle production. Surfactants are usually used as additives to enhance the production of microbial products by improving their secretion through the cell membrane.[134] The concentration of the electron shuttle directly affects the electron transfer efficiency between the bacteria and the electrode. Thus, surfactant addition is expected to enhance electron shuttle production and hence improve EET and MES performance. As determined by cyclic voltammetry (CV) and UV-Vis spectroscopy, the CV and UV-Vis peaks corresponding to the electron shuttle were significantly increased by EDTA or PEI addition in a *P. aeruginosa*-inoculated MFC.[140] The results implied that the concentration of the electron shuttle secreted by the microorganism was enhanced by the addition of surfactants. In another *P. aeruginosa*-inoculated air cathode MFC, SOP addition significantly increased the CV peaks corresponding to the electron shuttle (pyocyanin). Further analysis with HPLC showed that pyocyanin (a kind of phenazine electron shuttle) secreted by *P. aeruginosa* was increased from ~7 mg L^{-1} to ~14.5 mg L^{-1}. The increased pyocyanin production directly resulted in ~4 times enhancement of the power output.[131] It is supposed that the increased cell permeability through surfactant addition would directly promote the secretion of the electron shuttle, which might relieve the intracellular product feedback inhibition and thus enhance the flux of the electron shuttle synthesis pathway. However, the detailed mechanism should be further explored.

5.4.2   Metal Ions Improve Cell Adherence for Efficient MES

Because metal ions play essential roles in various biological processes, such as cell growth, enzyme catalysis, and signal transduction, modulation of metal ions was considered to be effective to manipulate cell physiology and thus cell metabolism. The use of metal ions as inducers for the improvement of secondary metabolism production from microorganisms has been extensively studied and was demonstrated to be a practical and promising chemical bioengineering strategy.[128,144]

Iron was successfully used as a chemical inducer to improve the performance of MES.[145] As *Shewanella* is an iron-reducing bacteria, the availability of iron in the medium significantly affected its physiology and electron transfer pathway. For example, the presence of poorly soluble Fe(III) compounds promoted the synthesis and secretion of the electron shuttle (flavins) production.[146] Wu *et al.* found that addition of ferric citrate (6 mM) at the start-up stage or after the biofilm establishment on the electrode would result in about a 1-fold increase of the power output in a MFC system.[145] Further analysis showed that the flavins secreted by the bacteria increased from ~3.5 μM per g protein to ~5 μM per g protein upon Fe(III) addition. The increased accumulation of the electron shuttle directly enhanced the EET between the cells and the electrode, which largely contributed to the improvement in the power output.[145] Moreover, as compared with the control, Fe(III) addition also increased the biomass loading on the electrode biofilm, which should be another important factor in the promotion of the energy output.[145] Taken together, Fe(III) addition induced flavins secretion and biofilm development by *S. oneidensis* MR-1, and facilitated the EET between cells and the electrode, which eventually improved the performance of MES.[145] It was reported that iron also acted as a signaling molecule that regulated the *Pseudomonas* quinolone signal (PQS) system and iron acquisition system, and hence affected the biofilm development.[147] However, how the iron affected the biofilm development and flavin secretion/synthesis in *Shewanella* is still unclear and deserves further investigation.

Calcium is another interesting metal ion that is extensively used as a chemical inducer to modulate the secondary metabolism of microorganisms.[144] It was also effectively applied in a MES system for performance improvement. By supplementing 1.4 mM CaCl$_2$ into the electrolyte, the *S. oneidensis* MR-1 inoculated MFC delivered over 80% higher current output and ~1 fold higher maximum power output than that of the control.[148] Further analysis showed that addition of CaCl$_2$ at that concentration had little influence on the bulk conductivity and internal resistance of the MFC system, but it induced mild cell aggregation of *S. oneidensis* MR-1. The authors suggested that the aggregates induced by CaCl$_2$ are more stable than cells in suspension and thus are able to produce a higher power output.[148] However, the relationship between the aggregation and increased power output needs to be further validated. Moreover, as the calcium ion is a widespread secondary messenger in bacterial signal transduction networks,[144] exploration of its effects on signal transduction and its relationship with the EET and power output might be helpful to gain new insights into this chemical bioengineering process.

5.4.3 Quorum Sensing Signaling-Mediated Population Manipulation for Efficient MES

Quorum sensing is a widespread cell–cell communication system in microorganisms in which cells use small-molecule chemicals as signaling molecules to induce population-level responses.[149,150] With quorum sensing systems, bacteria can sense their population density and coordinate their behaviors at the community/population level, which ensures that the community completes the complex/sophisticated processes that individuals are incapable of.[150] Thus, quorum sensing regulation is a typical and natural chemical bioengineering process that uses endogenously produced chemicals to modulate biological reactions/processes/behaviors. In MES, performance improvement by manipulation of the quorum sensing system in *P. aeruginosa* was successfully demonstrated.[151]

Overproduction of quorum sensing signaling molecules was attempted to modulate the EET and performance of MES.[34] As *rhlI* is directly responsible for *N*-butyryl homoserine lactone (BHL, a quorum sensing signaling molecule in *P. aeruginosa*) synthesis, Yong *et al.* overproduced BHL by overexpression of the *rhlI/rhlR* gene cluster in *P. aeruginosa* CGMCC1.860.[34,152,153] The BHL signaling molecule binds with the RhlR regulator protein, and then activates the synthesis of phenazine (the electron shuttle for *P. aeruginosa*). Therefore, the bacteria with enhanced BHL production showed much higher electrochemical activity and higher phenazine production than the control.[34] Moreover, the energy output of the MES system inoculated with the engineered *P. aeruginosa* strain was increased about 1.6 times over the system inoculated with the control strain.[34]

Furthermore, indirect manipulation of the *rhl* quorum sensing system was also explored to improve the performance of MES.[154] For example, RetS was proved to be a global regulator and indirectly repressed the *rhl* quorum sensing system through a GacS/GacA signaling cascade in *P. aeruginosa*.[155] Thus, disruption of the *retS* gene was expected to relieve the expression of *gacS/gacA* and in turn activate the expression of the *rhl* quorum sensing system.[154] By disrupting the *retS* gene, the *P. aeruginosa* mutant generated a ~45 fold higher current than that generated by the wild-type strain. Moreover, extraneously supplementing with homoserine lactones (quorum sensing molecules) directly induced a ~28-fold increase in the current output.[154] PQS is another quorum sensing system of *P. aeruginosa* that indirectly regulates phenazine synthesis *via* the 2-heptyl-3-hydroxy-4-quinolone signaling molecule.[156,157] Wang *et al.* genetically engineered the PQS system by overexpressing *pqsE* and disruption of *pqsC*, which resulted in a ~1 fold increase in phenazine production and lead to a ~5 fold improvement in the current output from a *P. aeruginosa*-inoculated MFC.[158]

In consideration of the ubiquity of quorum sensing systems in microorganisms, the use of quorum sensing signaling molecules or their analogues as chemical inducers to modulate the EET and performance of MES could be a promising chemical bioengineering strategy.

5.4.4 Cell Immobilization for Improved MES

As the electrocatalysis of MES occurs at the interface between the bacteria and the solid electrode, the cells in suspension in the bulk electrolyte are inefficient in electrocatalysis due to the long-range EET route and high energy demand for rapid growth and cell maintenance. Thus, shortening the EET route and reducing the energy consumption of microorganisms is essential for MES to achieve high efficiency. Biofilm adhered to the electrode plays a determinant role in EET *via* direct contact-based electron transfer (redox membrane protein/conductive pili mediated) and electron shuttle mediated electron transfer.[26,39] Electrode-attached biofilms can facilitate EET due to the short EET transfer route and high local cell density. However, naturally formed biofilms are usually non-conductive due to the encapsulation by EPS and the insulation property of the cell membrane, which limits the bacterial EET and electrocatalysis.[26] Cell immobilization is an efficient strategy to form an artificial biofilm on a substrate. Nevertheless, conventional cell immobilization methods are normally used to encapsulate cells into non-conductive polymers, such as agar, chitosan, and alginate.[159] Thus, these methods cannot be directly used for electrogen immobilization as these polymers may limit the EET. Our group attempted to develop new strategies for electrogen immobilization with the aim of maintaining the electroactivity of the electrogens.[26,39,159] A simple strategy based on the most extensively used alginate immobilization method was developed, *i.e.*, the electrogenic cells were mixed with conductive graphite particles, and then immobilized with alginate to form conductive graphite/alginate granules.[159] Furthermore, the conductive polymer polypyrrole (PPy) was employed as an alternative to alginate to further improve the conductivity of the immobilized electrogens matrix. By using this method, an artificial conductive biofilm was developed on the solid electrode, and showed a dramatic improvement in the electrocatalysis activity.[26] More interestingly, we designed a self-immobilization process to form a 3D macroporous conductive biofilm by using graphene oxide (GO) as the supporting substrate.[39] Without using any polymer, the water soluble GO (just like a fishing net) could immobilize the bacterial cells (fish) on, or wrapped into, the nanosheets. Surprisingly, the GO nanosheets with low conductivity were spontaneously reduced to highly conductive graphene and self-assembled into a 3D macroporous bacteria/graphene network. This 3D network not only retained the electroactivity of the electrogenic bacteria, but also reinforced the EET between the bacteria and the electrode owing to the excellent conductivity of the graphene and enhanced electric interaction between the cells and the graphene.[39] Thus, the self-assembly strategy opened a new dimension for cell immobilization of electrogenic bacteria.

The improvement in MES performance by the cell immobilization strategy is mainly ascribed to the increased biomass loading, biofilm conductivity, and electric interaction.[26,39] According to the lifecycle of the biofilm, the bacterial cells will detach from the matrix after the maturation of the natural biofilm, which limits the biomass loading and thickness of the biofilm. By using *S. oneidensis* MR-1 as the model strain, the naturally occurring biofilm on the electrode usually has a thickness less than 10 μm, while the artificial conductive biofilm constructed by us can reach a thickness up to 600 μm and the thickness of the graphene hybrid biofilm reached an impressive ~4 mm.[26,39] More importantly, the increased conductivity of the synthetic conductive biofilm significantly reduced the internal resistance of the MES, *e.g.*, the MES with artificial conductive biofilm (~3.2 mS cm^{-1}) showed about 10 times lower ohmic resistance than that with the naturally occurring biofilm (4.6 kΩ *vs.* 49 kΩ).[26] By packing with the conductive graphite/alginate/cells granules, the charge transfer resistance of the MES system decreased from ~2700 Ω (suspension cells without immobilization) to ~1600 Ω.[159] Strikingly, the electroactivity of the synthetic/hybrid conductive biofilm was dramatically improved as the immobilization process might artificially/spontaneously reinforce the interaction between the cells and the nanostructured conductive materials.[39] For example, the naturally occurring biofilm of *S. oneidensis* MR-1 usually shows asymmetric or even undetectable CV peaks from c-type cytochrome proteins (outer membrane proteins, responsible for contact-based direct EET), which indicates hindered EET between these proteins and solid electrode.[38,39,160,161] However, the graphene/bacteria hybrid biofilm showed clear and symmetric CV peaks of c-type cytochrome proteins, which indicates that the cell immobilization process facilitates the direct-contact based EET between cells and the electrode.[38]

In addition, cell immobilization also increased cell tolerance to environmental stress and fluctuating conditions. By immobilizing *S. oneidensis* MR-1 in the conductive graphite/alginate granules, the MES system packed with these granules showed much higher tolerance to the shock of high salt concentration than the system with the cells in suspension.[159] For example, upon addition of a high concentration of NaCl, the voltage of the MFC with cells in suspension decreased ~60%, while only a ~19% decrease was observed for the MFC with immobilized cells. More impressively, the voltage output of the cell immobilization system was then quickly restored, while the suspension system is incapable of restoring the voltage output.[159] The immobilization system also showed much more stable performance in a continuous system due to the avoidance of cell washout.[162] Besides, cell immobilization also increased the coulombic efficiency of the MFCs, possibly due to reduced cell growth. The coulombic efficiency obtained from the MFCs with graphite/

alginate granules immobilized cells is about 0.6–1.8 times higher than with cells in suspension.[159] From the practical application viewpoint, immobilization is very important for decreasing sludge production and the ease of re-usage. The immobilized biofilm is much more suitable for use in continuous flow systems owing to the biofilm stability, tolerance to culture condition fluctuations, reduction of cell washout, and enhanced energy efficiency.

5.5 Concluding Remarks and Perspectives

This chapter reviewed the recent progress in the manipulation of MES systems with chemical strategies and highlighted the application of "chemical bioengineering" in this field. It introduced the chemistry of MES, including MFC and MESy systems, with emphasis on extracellular electron transfer, and summarized the recent endeavors in chemical manipulation of extracellular electron transfer and modulation of cell physiology in MES with various chemical strategies for performance enhancement. The great progress achieved with chemical modulation demonstrates that "chemical bioengineering", which refers to the manipulation of biological processes with chemical strategies, holds great promise in the field of MES.

However, the MES systems, especially for the newly developed processes (*e.g.*, MESy), are in the early stage of development and deeper understanding of the mechanisms is urgently required to achieve the technical breakthroughs that are essential to advance the application of these technologies. Now it is time to explore the detailed chemical and biological mechanisms, and special attention should be paid to the intracellular electron releasing process, bidirectional electron transfer between cells and the electrode, and the chemistry in the electron transfer process. Moreover, mechanism-directed approaches for optimization of the energy efficiency of MES should be developed. For practical applications, cost considerations including materials and operation costs, scale-up for large-scale applications as well as scale-down for miniature devices, and in-field testing should be taken into account. By taking the unique advantages of interdisciplinary research and collaboration, chemical biotechnology and bioengineering (which has already shown its great potential in advancing MFC technology and the emerging MES technologies) can be reasonably expected to accelerate the mechanistic understandings as well as the practical applications of MES.

Acknowledgements

We gratefully acknowledge the financial support from the National Natural Science Foundation of China (NSFC 21306069), Specialized Research Fund for the Doctoral Program of Higher Education (Ministry of Education, 20133227120014), Natural Science Foundation of Jiangsu Province (BK20130492), and 'Six Talent Peaks' program (2012-NY-029, Jiangsu Province, China).

References

1. B. G. Pollet, I. Staffell and J. L. Shang, *Electrochim. Acta*, 2012, **84**, 235–249.
2. Y. Xiang, S. F. Lu and S. P. Jiang, *Chem. Soc. Rev.*, 2012, **41**, 7291–7321.
3. V. S. Bagotsky, *J. Solid State Electrochem.*, 2011, **15**, 1559–1562.
4. C. G. Morales-Guio, L. A. Stern and X. L. Hu, *Chem. Soc. Rev.*, 2014, **43**, 6555–6569.
5. M. A. Rodrigo, N. Oturan and M. A. Oturan, *Chem. Rev.*, 2014, **114**, 8720–8745.
6. P. C. Hallenbeck, M. Grogger and D. Veverka, *Electrocatalysis*, 2014, **5**, 319–329.
7. S. V. Mohan, G. Velvizhi, K. V. Krishna and M. L. Babu, *Bioresour. Technol.*, 2014, **165**, 355–364.
8. F. Zhao, R. C. T. Slade and J. R. Varcoe, *Chem. Soc. Rev.*, 2009, **38**, 1926–1939.
9. B. E. Logan and K. Rabaey, *Science*, 2012, **337**, 686–690.
10. D. Call and B. E. Logan, *Environ. Sci. Technol.*, 2008, **42**, 3401–3406.
11. K. Rabaey and R. A. Rozendal, *Nat. Rev. Microbiol.*, 2010, **8**, 706–716.
12. B. E. Logan and M. Elimelech, *Nature*, 2012, **488**, 313–319.
13. R. D. Cusick, Y. Kim and B. E. Logan, *Science*, 2012, **335**, 1474–1477.
14. M. A. Arugula, N. Shroff, E. Katz and Z. He, *Chem. Commun.*, 2012, **48**, 10174–10176.
15. B. E. Logan, *Nat. Rev. Microbiol.*, 2009, **7**, 375–381.
16. D. R. Lovley, *Nat. Rev. Microbiol.*, 2006, **4**, 497–508.
17. J. B. Arends and W. Verstraete, *Microb. Biotechnol.*, 2012, **5**, 333–346.
18. M. C. Potter, *Proc. R. Soc. London, Ser. B*, 1911, **84**, 260–276.
19. V. B. Oliveira, M. Simoes, L. F. Melo and A. M. F. R. Pinto, *Biochem. Eng. J.*, 2013, **73**, 53–64.
20. J. Khera and A. Chandra, *Proc. Natl. Acad. Sci., India, Sect. A*, 2012, **82**, 31–41.
21. J. Liu, Y. C. Yong, H. Song and C. M. Li, *ACS Catal.*, 2012, **2**, 1749–1752.
22. Y. C. Yong, Y. Y. Yu, Y. Yang, C. M. Li, R. R. Jiang, X. Wang, J. Y. Wang and H. Song, *Electrochem. Commun.*, 2012, **19**, 13–16.
23. Y. C. Yong and J. Z. Sun, *Industrial Biotechnology Development Report*, 2013.
24. Y. G. Yang, M. Y. Xu, J. Guo and G. P. Sun, *Process Biochem.*, 2012, **47**, 1707–1714.
25. D. R. Bond, D. E. Holmes, L. M. Tender and D. R. Lovley, *Science*, 2002, **295**, 483–485.
26. Y. Y. Yu, H. L. Chen, Y. C. Yong, D. H. Kim and H. Song, *Chem. Commun.*, 2011, **47**, 12825–12827.
27. M. Lu and S. F. Y. Li, *Crit. Rev. Environ. Sci. Technol.*, 2012, **42**, 2504–2525.
28. X. Y. Yong, J. Feng, Y. L. Chen, D. Y. Shi, Y. S. Xu, J. Zhou, S. Y. Wang, L. Xu, Y. C. Yong, Y. M. Sun, C. L. Shi, P. K. OuYang and T. Zheng, *Biosens. Bioelectron.*, 2014, **56**, 19–25.

29. C. Shea, P. Clauwaert, W. Verstraete and R. Nerenberg, *Water Sci. Technol.*, 2008, **58**, 1941–1946.
30. Y. Du, Y. J. Feng, Y. P. Qu, J. Liu, N. Q. Ren and H. Liu, *Environ. Sci. Technol.*, 2014, **48**, 7634–7641.
31. S. Kondaveeti and B. Min, *Bioprocess Biosyst. Eng.*, 2013, **36**, 231–238.
32. S. Pandit, B. K. Nayak and D. Das, *Bioresour. Technol.*, 2012, **107**, 97–102.
33. Z. Du, H. Li and T. Gu, *Biotechnol. Adv.*, 2007, **25**, 464–482.
34. Y. C. Yong, Y. Y. Yu, C. M. Li, J. J. Zhong and H. Song, *Biosens. Bioelectron.*, 2011, **30**, 87–92.
35. A. P. Borole, D. Aaron, C. Y. Hamilton and C. Tsouris, *Environ. Sci. Technol.*, 2010, **44**, 2740–2744.
36. A. Sikora, J. Wojtowicz-Sienko, P. Piela, U. Zielenkiewicz, K. Tomczyk-Zak, A. Chojnacka, R. Sikora, P. Kowalczyk, E. Grzesiuk and M. Blaszczyk, *J. Microbiol. Biotechnol.*, 2011, **21**, 305–316.
37. X. Xia, Y. M. Sun, P. Liang and X. Huang, *Bioresour. Technol.*, 2012, **120**, 26–33.
38. Y. C. Yong, X. C. Dong, M. B. Chan-Park, H. Song and P. Chen, *ACS Nano*, 2012, **6**, 2394–2400.
39. Y. C. Yong, Y. Y. Yu, X. H. Zhang and H. Song, *Angew. Chem., Int. Ed.*, 2014, **53**, 4480–4483.
40. J. Liu, Y. Qiao, C. X. Guo, S. Lim, H. Song and C. M. Li, *Bioresour. Technol.*, 2012, **114**, 275–280.
41. B. Y. Jia, D. W. Hu, B. Z. Xie, K. Dong and H. Liu, *Biosens. Bioelectron.*, 2013, **41**, 894–897.
42. I. A. Ieropoulos, J. Greenman and C. Melhuish, *Int. J. Hydrogen Energy*, 2013, **38**, 492–496.
43. Y. Z. Fan, H. Q. Hu and H. Liu, *Environ. Sci. Technol.*, 2007, **41**, 8154–8158.
44. K. P. Nevin, T. L. Woodard, A. E. Franks, Z. M. Summers and D. R. Lovley, *mBio*, 2010, **1**, e00103–e00110.
45. A. V. Pandit and R. Mahadevan, *Microb. Cell Fact.*, 2011, **10**, 76.
46. D. R. Lovley and K. P. Nevin, *Curr. Opin. Biotechnol.*, 2013, **24**, 385–390.
47. D. E. Ross, J. M. Flynn, D. B. Baron, J. A. Gralnick and D. R. Bond, *PLoS One*, 2011, **6**, e16649.
48. K. Rabaey, P. Girguis and L. K. Nielsen, *Curr. Opin. Biotechnol.*, 2011, **22**, 371–377.
49. L. P. Huang, J. M. Regan and X. Quan, *Bioresour. Technol.*, 2011, **102**, 316–323.
50. C. W. Marshall, D. E. Ross, E. B. Fichot, R. S. Norman and H. D. May, *Appl. Environ. Microbiol.*, 2012, **78**, 8412–8420.
51. K. P. Nevin, S. A. Hensley, A. E. Franks, Z. M. Summers, J. H. Ou, T. L. Woodard, O. L. Snoeyenbos-West and D. R. Lovley, *Appl. Environ. Microbiol.*, 2011, 77, 2882–2886.
52. H. R. Nie, T. Zhang, M. M. Cui, H. Y. Lu, D. R. Lovley and T. P. Russell, *Phys. Chem. Chem. Phys.*, 2013, **15**, 14290–14294.
53. V. Flexer, J. Chen, B. C. Donose, P. Sherrell, G. G. Wallace and J. Keller, *Energy Environ. Sci.*, 2013, **6**, 1291–1298.

54. T. Zhang, H. R. Nie, T. S. Bain, H. Y. Lu, M. M. Cui, O. L. Snoeyenbos-West, A. E. Franks, K. P. Nevin, T. P. Russell and D. R. Lovley, *Energy Environ. Sci.*, 2013, **6**, 217–224.

55. C. W. Marshall, D. E. Ross, E. B. Fichot, R. S. Norman and H. D. May, *Environ. Sci. Technol.*, 2013, **47**, 6023–6029.

56. K. J. J. Steinbusch, H. V. M. Hamelers, J. D. Schaap, C. Kampman and C. J. N. Buisman, *Environ. Sci. Technol.*, 2010, **44**, 513–517.

57. H. Li, P. H. Opgenorth, D. G. Wernick, S. Rogers, T. Y. Wu, W. Higashide, P. Malati, Y. X. Huo, K. M. Cho and J. C. Liao, *Science*, 2012, **335**, 1596–1596.

58. E. Marsili, D. B. Baron, I. D. Shikhare, D. Coursolle, J. A. Gralnick and D. R. Bond, *Proc. Natl. Acad. Sci. U. S. A.*, 2008, **105**, 3968–3973.

59. K. Rabaey, N. Boon, M. Hofte and W. Verstraete, *Environ. Sci. Technol.*, 2005, **39**, 3401–3408.

60. D. H. Park and J. G. Zeikus, *Appl. Environ. Microbiol.*, 2000, **66**, 1292–1297.

61. Y. Qiao, S. J. Bao, C. M. Li, X. Q. Cui, Z. S. Lu and J. Guo, *ACS Nano*, 2008, **2**, 113–119.

62. F. Aulenta, A. Catervi, M. Majone, S. Panero, P. Reale and S. Rossetti, *Environ. Sci. Technol.*, 2007, **41**, 2554–2559.

63. R. S. Hartshorne, C. L. Reardon, D. Ross, J. Nuester, T. A. Clarke, A. J. Gates, P. C. Mills, J. K. Fredrickson, J. M. Zachara, L. Shi, A. S. Beliaev, M. J. Marshall, M. Tien, S. Brantley, J. N. Butt and D. J. Richardson, *Proc. Natl. Acad. Sci. U. S. A.*, 2009, **106**, 22169–22174.

64. R. Nakamura, K. Ishii and K. Hashimoto, *Angew. Chem., Int. Ed.*, 2009, **48**, 1606–1608.

65. A. Okamoto, K. Hashimoto and K. H. Nealson, *Angew. Chem., Int. Ed.*, 2014, **53**, 10988–10991.

66. M. Y. El-Naggar, G. Wanger, K. M. Leung, T. D. Yuzvinsky, G. Southam, J. Yang, W. M. Lau, K. H. Nealson and Y. A. Gorby, *Proc. Natl. Acad. Sci. U. S. A.*, 2010, **107**, 18127–18131.

67. G. Reguera, K. D. McCarthy, T. Mehta, J. S. Nicoll, M. T. Tuominen and D. R. Lovley, *Nature*, 2005, **435**, 1098–1101.

68. X. C. Jiang, J. S. Hu, L. A. Fitzgerald, J. C. Biffinger, P. Xie, B. R. Ringeisen and C. M. Lieber, *Proc. Natl. Acad. Sci. U. S. A.*, 2010, **107**, 16806–16810.

69. N. S. Malvankar, M. Vargas, K. P. Nevin, A. E. Franks, C. Leang, B. C. Kim, K. Inoue, T. Mester, S. F. Covalla, J. P. Johnson, V. M. Rotello, M. T. Tuominen and D. R. Lovley, *Nat. Nanotechnol.*, 2011, **6**, 573–579.

70. N. S. Malvankar, S. E. Yalcin, M. T. Tuominen and D. R. Lovley, *Nat. Nanotechnol.*, 2014, **9**, 1012–1017.

71. L. V. Richter, S. J. Sandler and R. M. Weis, *J. Bacteriol.*, 2012, **194**, 2551–2563.

72. R. M. Snider, S. M. Strycharz-Glaven, S. D. Tsoi, J. S. Erickson and L. M. Tender, *Proc. Natl. Acad. Sci. U. S. A.*, 2012, **109**, 15467–15472.

73. J. J. Chen, W. Chen, H. He, D. B. Li, W. W. Li, L. Xiong and H. Q. Yu, *Environ. Sci. Technol.*, 2013, **47**, 1033–1039.

74. X. W. Liu, X. F. Sun, J. J. Chen, Y. X. Huang, J. F. Xie, W. W. Li, G. P. Sheng, Y. Y. Zhang, F. Zhao, R. Lu and H. Q. Yu, *Sci. Rep.*, 2013, **3**, 1616.
75. D. Coursolle, D. B. Baron, D. R. Bond and J. A. Gralnick, *J. Bacteriol.*, 2010, **192**, 467–474.
76. J. C. Biffinger, J. Pietron, O. Bretschger, L. J. Nadeau, G. R. Johnson, C. C. Williams, K. H. Nealson and B. R. Ringeisen, *Biosens. Bioelectron.*, 2008, **24**, 900–905.
77. Y. C. Yong, Z. Cai, Y. Y. Yu, P. Chen, R. R. Jiang, B. Cao, J. Z. Sun, J. Y. Wang and H. Song, *Bioresour. Technol.*, 2013, **130**, 763–768.
78. V. B. Wang, S. L. Chua, B. Cao, T. Seviour, V. J. Nesatyy, E. Marsili, S. Kjelleberg, M. Givskov, T. Tolker-Nielsen, H. Song, J. Say, C. Loo and L. Yang, *PLoS One*, 2013, **8**, e63129.
79. X. Y. Yong, D. Y. Shi, Y. L. Chen, J. Feng, L. Xu, J. Zhou, S. Y. Wang, Y. C. Yong, Y. M. Sun, P. K. OuYang and T. Zheng, *Bioresour. Technol.*, 2014, **152**, 220–224.
80. M. Izallalen, R. Mahadevan, A. Burgard, B. Postier, R. Didonato, J. Sun, C. H. Schilling and D. R. Lovley, *Metab. Eng.*, 2008, **10**, 267–275.
81. J. A. Smith, P. L. Tremblay, P. M. Shrestha, O. L. Snoeyenbos-West, A. E. Franks, K. P. Nevin and D. R. Lovley, *Appl. Environ. Microbiol.*, 2014, **80**, 4331–4340.
82. B. C. Kim, B. L. Postier, R. J. DiDonato, S. K. Chaudhuri, K. P. Nevin and D. R. Lovley, *Bioelectrochemistry*, 2008, **73**, 70–75.
83. T. H. Pham, N. Boon, K. De Maeyer, M. Hofte, K. Rabaey and W. Verstraete, *Appl. Microbiol. Biotechnol.*, 2008, **80**, 985–993.
84. S. B. Velasquez-Orta, I. M. Head, T. P. Curtis, K. Scott, J. R. Lloyd and H. von Canstein, *Appl. Microbiol. Biotechnol.*, 2010, **85**, 1373–1381.
85. A. Okamoto, R. Nakamura and K. Hashimoto, *Electrochim. Acta*, 2011, **56**, 5526–5531.
86. A. Okamoto, K. Hashimoto, K. H. Nealson and R. Nakamura, *Proc. Natl. Acad. Sci. U. S. A.*, 2013, **110**, 7856–7861.
87. R. A. Maithreepala and R. A. Doong, *J. Hazard. Mater.*, 2009, **164**, 337–344.
88. A. Okamoto, K. Saito, K. Inoue, K. H. Nealson, K. Hashimoto and R. Nakamura, *Energy Environ. Sci.*, 2014, **7**, 1357–1361.
89. G. Najafpour, M. Rahimnejad and A. Ghoreyshi, *Energy Sources, Part A*, 2011, **33**, 2239–2248.
90. S. Nasirahmadi and A. A. Safekordi, *Int. J. Environ. Sci. Technol.*, 2011, **8**, 823–830.
91. M. T. Khan, S. H. Duncan, A. J. M. Stams, J. M. van Dijl, H. J. Flint and H. J. M. Harmsen, *ISME J.*, 2012, **6**, 1578–1585.
92. M. Masuda, S. Freguia, Y. F. Wang, S. Tsujimura and K. Kano, *Bioelectrochemistry*, 2010, **78**, 173–175.
93. C. M. Eggleston, J. Voros, L. Shi, B. H. Lower, T. C. Droubay and P. J. S. Colberg, *Inorg. Chim. Acta*, 2008, **361**, 769–777.
94. M. Breuer, K. M. Rosso and J. Blumberger, *Proc. Natl. Acad. Sci. U. S. A.*, 2014, **111**, 611–616.

95. A. Okamoto, K. Hashimoto and R. Nakamura, *Bioelectrochemistry*, 2012, **85**, 61–65.

96. L. A. Fitzgerald, E. R. Petersen, R. I. Ray, B. J. Little, C. J. Cooper, E. C. Howard, B. R. Ringeisen and J. C. Biffinger, *Process Biochem.*, 2012, **47**, 170–174.

97. N. S. Malvankar and D. R. Lovley, *ChemSusChem*, 2012, **5**, 1039–1046.

98. E. Sokullu, I. M. Palabiyik, F. Onur and I. H. Boyaci, *Eng. Life Sci.*, 2010, **10**, 297–303.

99. V. Coman, T. Gustavsson, A. Finkelsteinas, C. von Wachenfeldt, C. Hagerhall and L. Gorton, *J. Am. Chem. Soc.*, 2009, **131**, 16171–16176.

100. S. Timur, B. Haghighi, J. Tkac, N. Pazarhoglu, A. Telefoncu and L. Gorton, *Bioelectrochemistry*, 2007, **71**, 38–45.

101. S. A. Patil, K. Hasan, D. Leech, C. Hagerhall and L. Gorton, *Chem. Commun.*, 2012, **48**, 10183–10185.

102. K. Hasan, S. A. Patil, D. Leech, C. Hagerhall and L. Gorton, *Biochem. Soc. Trans.*, 2012, **40**, 1330–1335.

103. Y. Yuan and S. Kim, *Bull. Korean Chem. Soc.*, 2008, **29**, 168–172.

104. E. R. Zhang and Y. C. Zhang, *Adv. Mater. Res.*, 2012, **396–398**, 1794–1798.

105. T. Baati, P. Horcajada, R. Gref, P. Couvreur and C. Serre, *J Pharm. Biomed. Anal.*, 2011, **56**, 758–762.

106. M. Ghasemi, W. R. W. Daud, N. Mokhtarian, A. Mayahi, M. Ismail, F. Anisi, M. Sedighi and J. Alam, *Int. J. Hydrogen Energy*, 2013, **38**, 9525–9532.

107. C. M. Ding, H. Liu, Y. Zhu, M. X. Wan and L. Jiang, *Energy Environ. Sci.*, 2012, **5**, 8517–8522.

108. G. G. Kumar, V. G. S. Sarathi and K. S. Nahm, *Biosens. Bioelectron.*, 2013, **43**, 461–475.

109. S. Timur, U. Anik, D. Odaci and L. Gorton, *Electrochem. Commun.*, 2007, **9**, 1810–1815.

110. C. H. Feng, L. Ma, F. B. Li, H. J. Mai, X. M. Lang and S. S. Fan, *Biosens. Bioelectron.*, 2010, **25**, 1516–1520.

111. J. J. Sun, H. Z. Zhao, Q. Z. Yang, J. Song and A. Xue, *Electrochim. Acta*, 2010, **55**, 3041–3047.

112. Y. Yuan, S. G. Zhou, Y. Liu and J. H. Tang, *Environ. Sci. Technol.*, 2013, **47**, 14525–14532.

113. S. A. Patil, S. Chigome, C. Hagerhall, N. Torto and L. Gorton, *Bioresour. Technol.*, 2013, **132**, 121–126.

114. Z. S. Lv, Y. F. Chen, H. C. Wei, F. S. Li, Y. Hu, C. H. Wei and C. H. Feng, *Electrochim. Acta*, 2013, **111**, 366–373.

115. M. Ghasemi, W. R. W. Daud, S. H. A. Hassan, S. E. Oh, M. Ismail, M. Rahimnejad and J. M. Jahim, *J. Alloys Compd.*, 2013, **580**, 245–255.

116. S. D. Minteer, *Top. Catal.*, 2012, **55**, 1157–1161.

117. Y. Yuan, J. Ahmed, L. H. Zhou, B. Zhao and S. Kim, *Biosens. Bioelectron.*, 2011, **27**, 106–112.

118. P. Liang, H. Y. Wang, X. Xia, X. Huang, Y. H. Mo, X. X. Cao and M. Z. Fan, *Biosens. Bioelectron.*, 2011, **26**, 3000–3004.

119. X. Xie, L. B. Hu, M. Pasta, G. F. Wells, D. S. Kong, C. S. Criddle and Y. Cui, *Nano Lett.*, 2011, **11**, 291–296.

120. Z. M. He, J. Liu, Y. Qiao, C. M. Li and T. T. Y. Tan, *Nano Lett.*, 2012, **12**, 4738–4741.

121. X. Gao, Y. Z. Zhang, X. W. Li and J. S. Ye, *Mater. Lett.*, 2013, **105**, 24–27.

122. J. X. Hou, Z. L. Liu and P. Y. Zhang, *J. Power Sources*, 2013, **224**, 139–144.

123. Y. Q. Wang, B. Li, D. Cui, X. D. Xiang and W. S. Li, *Biosens. Bioelectron.*, 2014, **51**, 349–355.

124. A. Mehdinia, E. Ziaei and A. Jabbari, *Electrochim. Acta*, 2014, **130**, 512–518.

125. I. H. Park, M. Christy, P. Kim and K. S. Nahm, *Biosens. Bioelectron.*, 2014, **58**, 75–80.

126. F.-F. Yan, Y.-R. He, C. Wu, Y.-Y. Cheng, W.-W. Li and H.-Q. Yu, *Environ. Sci. Technol. Lett.*, 2013, **1**, 128–132.

127. S. H. Roh, *J. Nanosci. Nanotechnol.*, 2013, **13**, 4158–4161.

128. Y. N. Xu, X. X. Xia and J. J. Zhong, *Biotechnol. Bioeng.*, 2014, **111**, 2358–2365.

129. Z. H. Wei, L. Q. Bai, Z. X. Deng and J. J. Zhong, *Bioresour. Technol.*, 2011, **102**, 1783–1787.

130. Y. C. Yong and J. J. Zhong, *Process Biochem.*, 2010, **45**, 1944–1948.

131. H. B. Shen, X. Y. Yong, Y. L. Chen, Z. H. Liao, R. W. Si, J. Zhou, S. Y. Wang, Y. C. Yong, P. K. OuYang and T. Zheng, *Bioresour. Technol.*, 2014, **167**, 490–494.

132. Y. C. Yong, Y. Y. Yu, Y. Yang, J. Liu, J. Y. Wang and H. Song, *Biotechnol. Bioeng.*, 2013, **110**, 408–416.

133. M. Wu, T. J. Yang, M. D. Miao and J. B. Ni, *Acta Chim. Sin.*, 2009, **67**, 2133–2138.

134. D. V. Cortez and I. C. Roberto, *Biocatal. Biotransform.*, 2014, **32**, 34–38.

135. S. C. Yang, M. Lei and T. B. Chen, *Soil Sediment Contam.*, 2014, **23**, 715–724.

136. A. Singh, J. D. Van Hamme and O. P. Ward, *Biotechnol. Adv.*, 2007, **25**, 99–121.

137. S. Paria, *Adv. Colloid Interface Sci.*, 2008, **138**, 24–58.

138. Q. Wen, F. Y. Kong, F. Ma, Y. M. Ren and Z. Pan, *J. Power Sources*, 2011, **196**, 899–904.

139. Q. Wen, F. Y. Kong, Y. M. Ren, D. X. Cao, G. L. Wang and H. T. Zheng, *Electrochem. Commun.*, 2010, **12**, 1710–1713.

140. J. Liu, Y. Qiao, Z. S. Lu, H. Song and C. M. Li, *Electrochem. Commun.*, 2012, **15**, 50–53.

141. S. Rebello, A. K. Asok, S. Mundayoor and M. S. Jisha, *Environ. Chem. Lett.*, 2014, **12**, 275–287.

142. E. Congiu and J. J. Ortega-Calvo, *Environ. Sci. Technol.*, 2014, **48**, 10869–10877.

143. K. K. S. Randhawa and P. K. S. M. Rahman, *Front. Microbiol.*, 2014, **5**, 454.

144. Y. N. Xu and J. J. Zhong, *Biotechnol. Adv.*, 2012, **30**, 1301–1308.

145. D. Wu, D. F. Xing, L. Lu, M. Wei, B. F. Liu and N. Q. Ren, *Bioresour. Technol.*, 2013, **135**, 630–634.

146. H. von Canstein, J. Ogawa, S. Shimizu and J. R. Lloyd, *Appl. Environ. Microbiol.*, 2008, **74**, 615–623.
147. A. G. Oglesby, J. M. Farrow, J. H. Lee, A. P. Tomaras, E. P. Greenberg, E. C. Pesci and M. L. Vasil, *J. Biol. Chem.*, 2008, **283**, 15558–15567.
148. L. A. Fitzgerald, E. R. Petersen, B. J. Gross, C. M. Soto, B. R. Ringeisen, M. Y. El-Naggar and J. C. Biffinger, *Biosens. Bioelectron.*, 2012, **31**, 492–498.
149. Y. C. Yong and J. J. Zhong, *Future Trends in Biotechnol.*, 2013, **131**, 25–61.
150. M. R. Parsek and E. P. Greenberg, *Trends Microbiol.*, 2005, **13**, 27–33.
151. Y. C. Yong, J. Z. sun, X. Y. Wu and H. Song, *Chemosphere*, 2014, In press.
152. Y. C. Yong and J. J. Zhong, *Bioresour. Technol.*, 2013, **136**, 761–765.
153. Y. C. Yong and J. J. Zhong, *Biosens. Bioelectron.*, 2009, **25**, 41–47.
154. A. Venkataraman, M. Rosenbaum, J. B. A. Arends, R. Halitschke and L. T. Angenent, *Electrochem. Commun.*, 2010, **12**, 459–462.
155. V. Venturi, *FEMS Microbiol. Rev.*, 2006, **30**, 274–291.
156. J. M. Farrow, Z. M. Sund, M. L. Ellison, D. S. Wade, J. P. Coleman and E. C. Pesci, *J. Bacteriol.*, 2008, **190**, 7043–7051.
157. M. Juhas, L. Eberl and B. Tummler, *Environ. Microbiol.*, 2005, **7**, 459–471.
158. V. B. Wang, S. L. Chua, B. Cao, T. Seviour, V. J. Nesatyy, E. Marsili, S. Kjelleberg, M. Givskov, T. Tolker-Nielsen, H. Song, J. S. Loo and L. Yang, *PLoS One*, 2013, **8**, e63129.
159. Y. C. Yong, Z. H. Liao, J. Z. Sun, T. Zheng, R. R. Jiang and H. Song, *Process Biochem.*, 2013, **48**, 1947–1951.
160. L. Peng, S. J. You and J. Y. Wang, *Biosens. Bioelectron.*, 2010, **25**, 1248–1251.
161. L. Peng, S. J. You and J. Y. Wang, *Biosens. Bioelectron.*, 2010, **25**, 2530–2533.
162. H. R. Luckarift, S. R. Sizemore, J. Roy, C. Lau, G. Gupta, P. Atanassov and G. R. Johnson, *Chem. Commun.*, 2010, **46**, 6048–6050.
163. A. Okamoto, K. Hashimoto, K. H. Nealson and R. Nakamura, *Proc. Natl. Acad. Sci. U. S. A.*, 2013, **110**, 7856–7861.
164. J. A. Smith, P. L. Tremblay, P. M. Shrestha, O. L. Snoeyenbos-West, A. E. Franks, K. P. Nevin and D. R. Lovley, *Appl. Environ. Microbiol.*, 2014, **80**, 4331–4340.
165. R. A. Maithreepala and R. A. Doong, *J. Hazard. Mater.*, 2009, **16**, 337–344.
166. C. H. Feng, L. Ma, F. B. Li, H. J. Mai, X. M. Lang and S. S. Fan, *Biosens. Bioelectron.*, 2010, **25**, 1516–1520.
167. S. Timur, U. Anik, D. Odaci and L. Gorton, *Electrochem. Commun.*, 2007, **9**, 1810–1815.
168. Y. Z. Zhang, G. Q. Mo, X. W. Li, W. D. Zhang, J. Q. Zhang, J. S. Ye, X. D. Huang and C. Z. Yu, *J. Power Sources*, 2011, **196**, 5402–5407.

CHAPTER 6

Chemical Bioengineering in Plant Cell Culture

FENGXIAN HU*[a], YUFANG XU[b], AND ZHENJIANG ZHAO[b]

[a]School of Biotechnology and State Key Laboratory of Bioreactor Engineering, East China University of Science and Technology, P.O. Box 283, 130 Meilong Road, Shanghai 200237, China; [b]Shanghai Key Laboratory of New Drug Design, School of Pharmacy, East China University of Science and Technology, Shanghai 200237, China
*E-mail: hufx@ecust.edu.cn

6.1 Plant Cell Culture

Plant cell culture technology originated from Cell Totipotency theory, which was proposed by the German botanist Haberlandt in 1902. The theory supposed that a single cultured plant cell containing full genetic information has the potential ability to regenerate the whole plant. This led to the possibility of *in vitro* plant cell culture. In 1922, the most successful cases of the early attempts involved the culture of maize, pea, and cotton root tips.[1,2] Tulecke achieved large-scale cultivation of rose stem cells for the first time in 1959.[3] Subsequently, plant cell culture technology was developed as a possible tool to produce plant secondary metabolites.

RSC Green Chemistry No. 34
Chemical Biotechnology and Bioengineering
By Xuhong Qian, Zhenjiang Zhao, Yufang Xu, Jianhe Xu, Y.-H. Percival Zhang, Jingyan Zhang, Yangchun Yong, and Fengxian Hu
© X.-H. Qian, Z.-J. Zhao, Y.-F. Xu, J.-H. Xu, Y.-H. P. Zhang, J.-Y. Zhang, Y.-C. Yong and F.-X. Hu, 2015
Published by the Royal Society of Chemistry, www.rsc.org

6.1.1   Plant Secondary Metabolites

Secondary metabolites (also called secondary products or natural compounds) are some small molecules that organisms produce to withstand adverse environments or physical adversity from pathogens, although they are not absolutely required for the survival of the organism. Secondary metabolites can be produced by many microbes, plants, fungi, and animals. Their synthesis pathways in organisms are shown in Figure 6.1.[4]

Among the worldwide total of 30 000 known natural products, about 80% stems from plant resources. The number of known chemical structures of plant secondary metabolites is four times the number of known microbial secondary metabolites. Plant secondary metabolites are widely used as valuable medicines (such as paclitaxel, vinblastine, camptothecin, ginsenosides, and artemisinin), food additives, flavors, spices (such as rose oil, vanillin), pigments (such as Sin red and anthocyanins), cosmetics (such as aloe polysaccharides), and bio-pesticides (such as pyrethrins).[5] Currently, a quarter of all prescribed pharmaceuticals compounds in industrialized countries are directly or indirectly derived from plants, or *via* semi-synthesis. Furthermore, 11% of the 252 drugs considered as basic and essential by the WHO are exclusively derived from plants.[6] According to their biosynthetic pathways, secondary metabolites are usually classified into three large molecule families: phenolics, terpenes, and steroids.[7] Some known plant-derived pharmaceuticals are shown in Table 6.1.

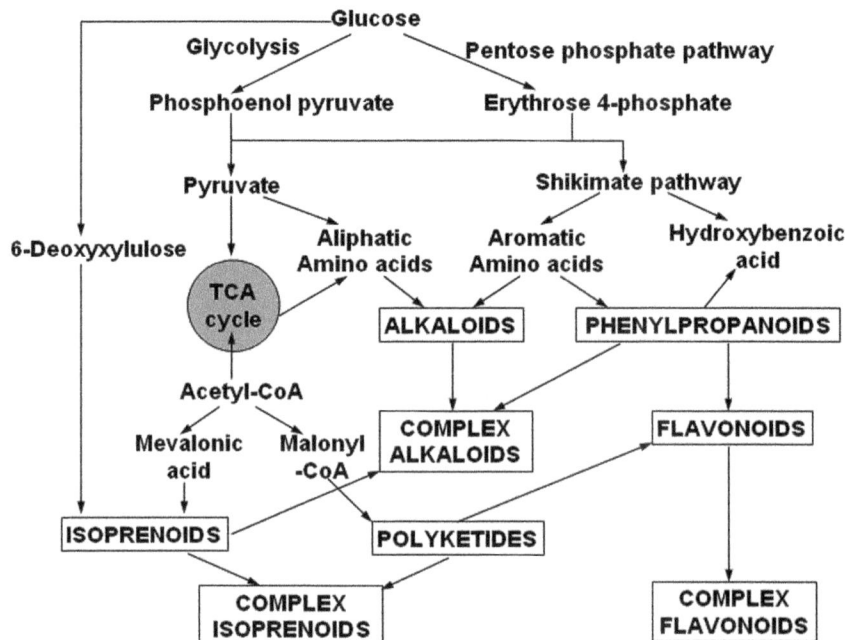

Figure 6.1   Biosynthetic relationship of major groups of secondary metabolites.[3]

6.1.2 Secondary Metabolites from Plant Cell Culture Technology

Plant cells in cultures can transform natural or artificial compounds through a variety of reactions, such as hydrogenation, dehydrogenation, isomerization, glycosylation, and hydroxylation, into secondary metabolite products. Moreover, the biosynthesis method is safe, controllable, and economical in contrast to field cultivation of their parent plants. In the past 50 years,

Table 6.1 Important pharmaceuticals from plant resources (adapted from ref. 8 and 9).

Product type	Plant species	Therapeutic use
Alkaloids		
Ajmalicine	*Catharanthus roseus*	Antihypertensive
Ajmaline	*Rauvolfia serpentina*	Antimalarial
Atropine, hyoscyamine, scopolamine	*Solanaceous*	Anticholinergic
Camptothecin	*Camptotheca acuminata* Decne	Antineoplastic
Capsaicin	*Capsicum*	Topical analgesic
Codeine, morphine	*Papaver somniferum*	Analgesic, antitussive
Cocaine	*Erythroxylum coca*	Local anaesthetic
Colchicine	*Colchicum autumnale*	Antigout
Ellipticine	*Ochrosia elliptica*	Antitumor
Emetine	*Cephaelis ipecacuanha*	Antiamoebic
Galanthamine	*Leucojum aestivum*	Cholinesterase inhibitor
Morphine	*Papaver somniferum*	Sedative
Nicotine	*Nicotiana*	Smoking cessation therapy
Physostigmine	*Physostigma venenosum*	Cholinergic
Pilocarpine	*Pilocarpus jaborandi*	Cholinergic
Quinine	*Cinchona*	Antimalarial
Quinidine	*Cinchona*	Cardiac depressant
Reserpine	*Rauwolfia serpentine*	Antihypertensive, psychotropic
Tubocurarine	*Chondodendron tomentosum, Strychnos toxifera*	Skeletal muscle relaxant
Vinblastine, vincristine	*C. roseus*	Antineoplastic
Yohimbine	Apocynaceae, Rubiaceae	Aphrodisiac
Terpenes and steroids		
Artemisinin	*Artemisia annua*	Antimalarial
Diosgenin, hecogenin, stigmasterol	*Dioscorea*	Oral contraceptives and hormonal drugs
Taxus brevifolia and other taxoids	*Taxus brevifolia*	Antineoplastic
Glycosides		
Digoxin, digitoxin	*Digitalis*	Cardiotonic
Sennosides A and B	*Cassia angustifolia*	Laxative
Others and mixtures		
Ipecac	*Cephaelis ipecacuanha*	Emetic
Podophyllotoxin	*Podophyllum peltatum*	Antineoplastic
Shikonin	*Lithospermum erythrorhizon*	Antibacterial

the application of plant cell culture technology for producing secondary metabolites has developed rapidly.[10] More than 400 plants have already been studied,[11] from which more than 600 kinds of secondary metabolites have been isolated from their cultured cells. To our satisfaction, the secondary metabolite products yield more than 60 kinds of plant cell cultures equaled or exceeded that of the corresponding original plants listed in Table 6.2.[12]

In 1983, shikonin was produced by plant cell cultures on an industrial scale for the first time from *Lithospermum erythrorhizon* cells by Mitsui Petrochemical Industries Ltd.[13] The success of shikonin production can be attributed to the selection of a high-product cell line, which accumulates ten-fold higher levels of shikonin than the roots of the mature plant. In the past several decades, a lot of effort has been put into plant cell, tissue, and organ culture as an alternative method to whole plant cultivation for the production of pharmacologically important plant secondary metabolites.[14] An application of plant cell culture technology has been commercialized production, as with *Coleus blumeicells* to produce rosmarinic acid and rosemary berberine (*Coptis japonica*) cells to produce berberine.[15] Recent advances in plant bioengineering have shown that bioreactor cultivation of adventitious roots is an attractive and alternative method to the whole plant, cell, or hairy root culture for biomass and bioactive compound production. Adventitious roots in phytohormone supplemented medium have been observed to have high rates of proliferation, tremendous potentialities of accumulation, and stable production of valuable secondary metabolites. Figure 6.2 shows the cultivation of adventitious roots in liquid-phase air-lift bioreactors (20–10 000 L).[16] A large number of studies have illustrated the feasibility of scaling up *in vitro* plant system-based processes while keeping their biosynthetic potential.[15]

Production of active compounds in plant cell culture can achieve commercial application depending on the following factors: the market price of the product, the market demand for the product, the cell growth rate, biomass production, and the activity of the compound. However, plant cell culture technology for natural active compound production is very limited at

Table 6.2　Secondary metabolite production from plant cell cultures compared with that from the parent plants.

Product	Plant	Yield (% DW)		Yield ratio (culture/plant)
		Culture	Plant	
Anthocyanin	*Vitis* sp.	16	10	1.6
	Euphorbia milii	4	0.3	13.3
	Perilla frutescens	24	1.5	16
Anthraquinone	*Morinda citrifolia*	18	2.2	8
Ajmalicine	*Catharanthus roseus*	1.0	0.3	3.3
Berberine	*Thalictrum minor*	10	0.1	1000
Ginsenoside	*Panax ginseng*	27	4.5	6
Nicotine	*Nicotiana tabacum*	3.4	2.0	1.7
Rosmarinic acid	*Lavandula vera* MM	13.4	0.2	67

the industrial level. For the low-yield issue of natural active compounds in plant cells, some researchers have explored deeply, including the breeding of high-quality cell lines, training, and the improvement of key genes cloned for synthetic active substances.[17] The low yield of natural active compounds from medicinal plant cells is still a "bottleneck".

The chemical synthesis of compounds from plant resources is often not economically feasible because of their highly complex structures and the specific stereo-chemical requirements of the compounds. The biotechnological production of valuable secondary metabolites in plant cells or organ cultures is an alternative method to the extraction of whole plant material. However, the contents of these compounds in plants using only plant cells or organ cultures are generally low, which limits their commercial application. If we extract the secondary metabolites directly from the naturally growing plants, especially for some of the rare plants, it can lead to depletion of plant resources. In addition, large-scale cultivation for the production of valuable secondary

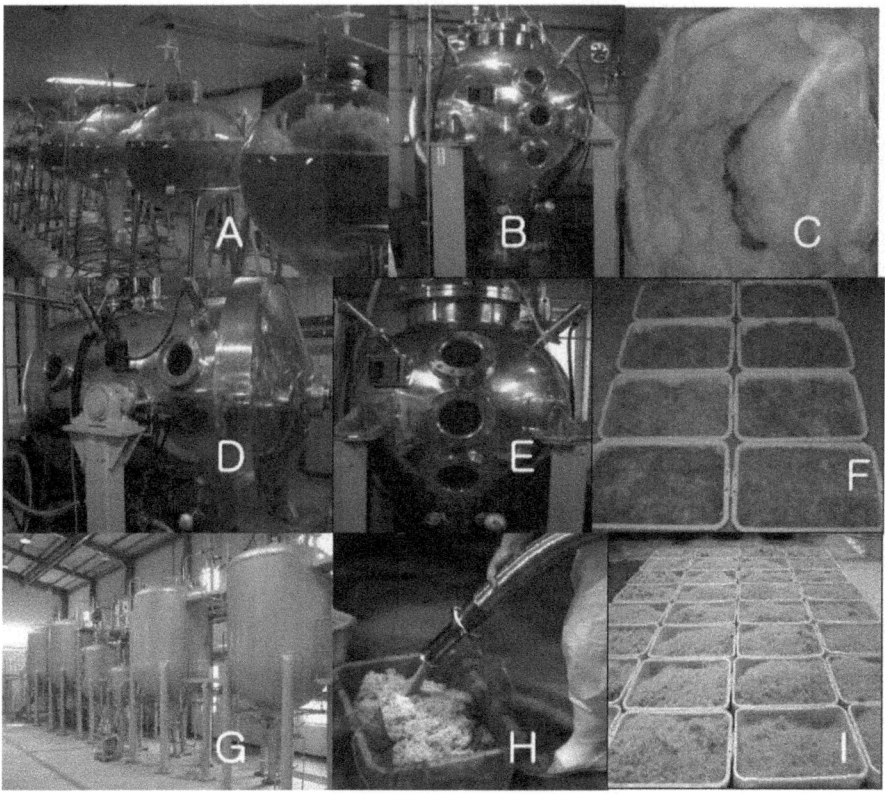

Figure 6.2 Cultivation of adventitious roots in liquid-phase air-lift bioreactors (20–10 000 L). Adventitious roots of *Echinacea purpurea*: 20 L, 500 L (A and B) and 1000 L (C and D); *Hypericum perforatum*: 500 L (E and F); *Ginseng*: 10 000 L (G and I).

metabolites in plants will occupy large tracts of arable land, and the quality and yield of secondary metabolites are also difficult to control. Therefore, it is necessary to find a better way to produce plant secondary metabolites.

6.1.3   Strategies to Improve Secondary Metabolite Production

Many biotechnological strategies have been taken and put into practice for improving the production of secondary metabolites from medicinal plants. These include: screening for high yielding cell lines, media modification, precursor feeding, and elicitation treatment.[18-23]

6.1.3.1   *Screening for High-Yielding Cell Lines*

The selection of cell lines includes the establishment of high-output and fast-growing cell line systems. It is possible to identify cell lines that can produce amounts of compounds equal or even higher than in the whole plant of origin.[24] Moreover, increasing metabolite levels by mutation breeding is another possible strategy, and selection of suitable analogues for this purpose could be an important factor in order to produce a variety of products.

6.1.3.2   *Optimizing Culture Conditions*

Plant cell culture medium includes inorganic compounds and phytohormones. Changing the medium components (concentration, proportion, and form) is a very powerful way of enhancing the culture efficiency of plant cell cultures. Unsuitable conditions often reduce the efficiency. For example, high auxin levels can stimulate cell growth, but can often negatively influence secondary metabolite production.[25]

6.1.3.3   *Feeding Precursors*

Feeding of precursors is based upon the idea that enhancing the concentrations of compounds that are intermediates or at the beginning of a biosynthetic route gives a good chance of promoting the yield of the final product. Attempts to increase the production of plant secondary metabolites by supplying precursors or intermediate compounds have been effective in many cases. In *Vanilla planifolia* cell culture, feeding ferulic acid led to an increase in vanillin accumulation.[26] Similarly, anthocyanin synthesis in *Daucus carota* was restored by the addition of dihydroquercetin (naringen).[27]

6.1.3.4   *Elicitation Treatment*

An elicitor may be defined as a substance that initiates plant cell stress responses and then affects cell metabolism processes when introduced in small concentrations to a living cell system. Elicitation is used to induce or enhance biosynthesis of secondary metabolites by the addition of trace amounts of elicitors.[28] Cell cultures have been established from many plants but the production of secondary metabolites is often too low for any practical application.[22,29] Therefore, in many cases, the production of secondary metabolites can be enhanced by treating the undifferentiated cells with elicitors, such as methyl jasmonate, salicylic acid, chitosan, and heavy metals.[28–31]

6.2 Elicitors in Plant Cell Culture

Since secondary metabolites protect plants from environmental changes, elicitation is the process of induced or enhanced synthesis of secondary metabolites by the plants to ensure their survival, persistence, and competitiveness. Thus, adding elicitors can selectively induce the expression of specific genes in plants to accumulate specific secondary metabolites.

In industry, enhancement of secondary metabolite production is crucial for successful application of plant cell culture technology.[32] Among the manipulative techniques available to promote the productivity of useful secondary metabolites from plant cell cultures, the treatment of plant cells with various elicitors has been one of the best approaches for dramatically increasing the production of useful secondary metabolites.[33–37]

6.2.1 Biotransformation Action Mechanism of Elicitors

The action mechanism of elicitors in biotransformation is not yet clear, but the initial hypothesis has been widely recognized. Specifically, elicitors act as external stimuli and bond to cell membrane receptors in plants, where they are identified and cause a series of reactions in the intracellular membrane, and then the activity and synthesis of secondary metabolite enzymes is regulated, ultimately leading to the synthesis and accumulation of secondary metabolites. The biotransformation action mechanism of elicitors is described in the following:

6.2.1.1 Membrane Changes

An elicitor is identified as an external signal to the cell, and can first bind to a cell membrane receptor. Further, the receptor–elicitor complex causes modifications in the membrane composition and permeability so as to cause the changes of the ion distribution. In recent years, a set of receptors has been confirmed by several researchers.[38,46] Most cell membrane elicitor receptors are summarized in Table 6.3. The receptor COI1 was recently identified as the JA receptor and accorded well with molecular modeling studies.[44,45]

6.2.1.2 Gene Expression

An existing enzyme may be activated by an elicitor, or the elicitor may switch on a gene for the mRNA transcription and translation of an enzyme. The translation of mRNA into proteins occurs through a series of events: initiation,

Table 6.3 Elicitors and their binding sites or receptors in plants.

Elicitor	Elicitor source	Elicitor properties	Plant receptors	References
N-Acetylchitoo-ligosaccharide	*Pyricularia oryzae*	Oligosaccharide	75 kDa in rice cells, 66 kDa in barley cells, 68 kDa in carrot cells, 60 kDa in wheat leaves	38 and 39
13-Pep	*Phytophthora soja*	13-Amino acid peptide from 42 kDa glycoprotein	19 kDa plasma membrane-bound protein with high binding affinity in parsley cell cultures	38
Harpins	*Pseudomonas* spp	10–12 kDa protein	115 kDa protein in microsomal tobacco or arabidopsis	39
Syringolides	*Escherichia coli* with *avir* gene	A glycolipid elicitor	34 kDa protein in soluble fraction from soy bean cell cultures	40
Cyptogeins	*Phytophthora cryptogea*; 10–12 kDa elicitins	Glycoprotein	193 kDa plasma membrane protein from tobacco	41
β-Glucan; chitin oligosaccharide	*Phytophthora sojae*, *Phytophthora megasperma*	1,6-β-Linked and 1,3-β-branched heptaglucoside	A receptor with h-glucan binding site and 1,3-β-glucanase activity in soybean cell cultures; 74 kDa, gene cloned in soybean root cells; 85 kDa membrane protein	42–44

elongation, and termination. Most of the regulatory mechanisms identified thus far mediate control of the initiation of mRNA translation, and therefore this stage and the signaling pathways that regulate it will be the primary focus of investigations into gene translation. In plant cell culture, it is reported that elicitors activate genes for mRNA transcription and translation in secondary metabolite production. In pea cell culture, MJA can induce cytochrome P450 (CYP93 family) gene expression.[47] Novel synthetic MJA derivatives also induce biosynthetic gene transcription and taxoid biosynthesis in cell cultures of *Taxus chinensis*.[48] In cell cultures of *Panax notoginseng*, gene transcription of ginsenoside biosynthesis was mediated by jasmonic acid in cell cultures of *Panax notoginseng* treated with chemically synthesized 2-hydroxyethyl jasmonate.[49]

6.2.1.3 Elicitor as the Secondary Messenger

Elicitors work as external signals on the cell membrane and can cause membrane changes, intracellular gene transcription promotion, and enzyme activity changes in a cascade process, which requires an intermediate medium to act as an intracellular messenger or secondary messenger.[50] JA and MJA are considered to be involved in the signal transduction pathway that induces particular enzymes to catalyze biochemical reactions to form low molecular weight defense compounds in plants, such as polyphenols, alkaloids, quinones, terpenoids, and polypeptides.[51-54]

6.2.2 Classification of Elicitors

The type and structure of elicitors vary greatly. Elicitors can be classified into abiotic elicitors or biotic elicitors on the basis of their "nature". The classification of elicitors is presented in Table 6.4. Abiotic elicitors are substances of non-biological origin, predominantly inorganic salts like Cu^{2+}, Cd^{2+}, and Ca^{2+}, as well as physical factors, osmotic stress,[55-57] synthetic organic compounds, and so on. Biotic elicitors are substances of biological origin. They

Table 6.4 Classification of elicitors in plant cell cultures.

Biotic elicitors	Abiotic elicitors
Directly released by microorganisms and recognized by the plant cell (enzymes, cell wall fragments)	UV light
Formed by action of microorganisms on plant cell wall (fragments of pectins *etc.*)	Extreme temperature (freezing, thawing)
Formed by the action of plant enzymes on microbial cell walls (chitosan, glucans)	High or low osmolarity
Compounds, endogenous and constitutive in nature, formed or released by the plant cell in response to various stimuli	Chemicals: heavy metals, nature chemical JA or SA, synthetic elicitors

include polysaccharides derived from plant cell walls (pectin or cellulose), micro-organisms (chitin or glucans), and glycoproteins or G-proteins or intracellular proteins whose functions are coupled to receptors and act by activating or inactivating a number of enzymes or ion channels.[58]

6.2.3 Natural Chemical Elicitors

6.2.3.1 Inorganic Metal Salts

Inorganic metal salts can induce secondary metabolism in plant cell cultures, such as Cu^{2+} and Ag^+. Cu^{2+} elicited phytoalexin production in rice plants.[59] Adding $AgNO_3$ induced taxol biosynthesis.[60] Rare earth metal salts, such as $(NH_4)_2Ce(NO_3)_6$, can also increase the accumulation of paclitaxel in *Taxus* cells.[61] Ginsenoside biosynthesis was elicited in cell cultures of *Panax ginseng* by vanadate.[62] The metabolic action of inorganic metal salts inducing the accumulation of secondary metabolites may be similar to other natural chemical elicitors.[63]

6.2.3.2 Organic JA and MJA

Since 1962, (–)-jasmonic acid methyl ester (–)-MJA has been well known as a fragrant constituent of the essential oils of *Jasminum* and other species.[64] The free acid (–)-JA was first isolated from the culture filtrate of a fungus in 1971.[65] The two substances attracted the attention of plant physiologists and were detected and identified as growth inhibitors in several higher plants.[66–68] Nowadays, JA and its stereoisomers (Figure 6.3) are the major representatives of a group of native plant regulators called jasmonates. They are widespread in the plant kingdom and show various physiological activities including growth, development, environmental responses in plants, and particularly defense responses against herbivores and necrotrophic pathogens. During the last decade, some studies have contributed to progress in understanding the biosynthesis and metabolism of natural JA analogues, as well as their physiological and molecular modes of action, which have been reviewed recently.[69]

Among the various elicitors, exogenous MJA has been confirmed as effective for the induction of secondary metabolites in plant cell cultures.[70] Mirjalili and Linden proved that paclitaxel content increased 15-fold after adding 10 μM MJA in *Taxus cuspdata*.[71] After treatment with MJA and ethylene together, paclitaxel production was increased 19-fold.[72] Zhong *et al.* demonstrated taxane production of 274 mg L^{-1} (23 days) and 527 mg L^{-1} (15 days) by adding sucrose or sucrose combined with methyl jasmonate, respectively.[90] It has also been established that treatment with exogenous MJA can elicit the accumulation of several classes of alkaloids, including the benzophenanthridines,[72] the vinca alkaloids,[73] and the tropane alkaloids.[74] Obviously, jasmonates are very significant elicitors that induce secondary metabolite production in plant cell cultures.[75]

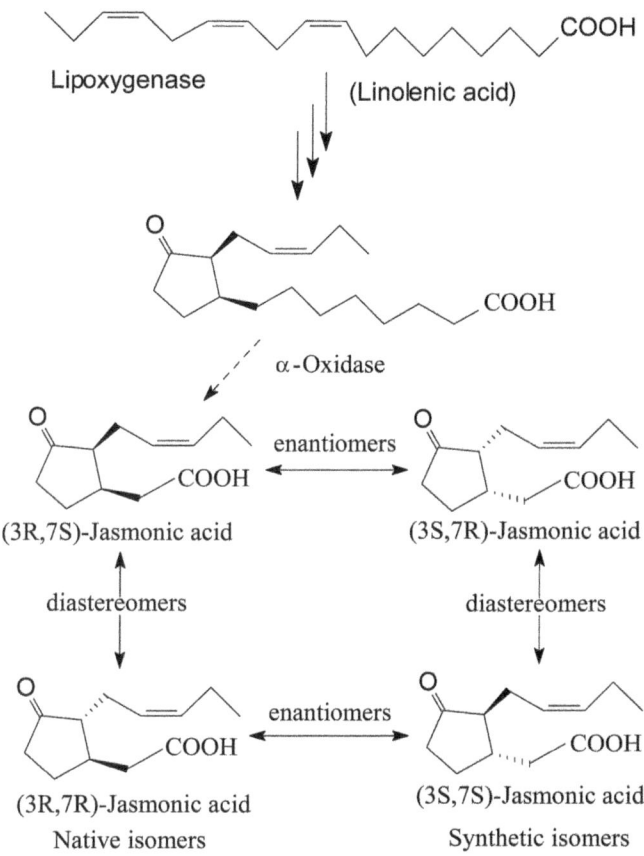

Figure 6.3 Metabolism of isomeric forms of jasmonic acid in plants.[58]

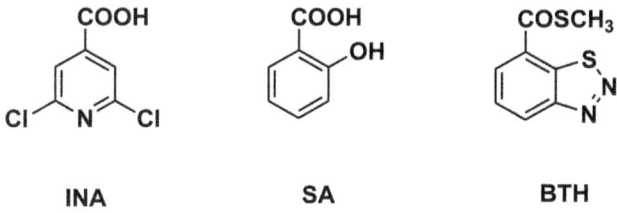

Figure 6.4 Chemical structures of elicitors INA, SA, and BTH.

6.2.3.3 Systemic Acquired Resistance (SAR) Activators

Besides jasmonates, some other compounds used for inducing systemic acquired resistance (SAR) in plants have now been found to be elicitors in plant cell cultures (Figure 6.4).

Salicylic acid (SA) is a phenolic compound biosynthesized through the phenylpropanoid pathway. SA has emerged as a key signaling component involved in the activation of certain plant defense responses.[76,77] Raskin proposed that SA was a new plant hormone,[78] but high concentrations of SA have a toxic effect on plants, so it has a narrow concentration range of security, which limits its further application in disease control. Exogenous SA can stimulate endogenous secondary metabolism in some plant species.[79,80] SA as an elicitor has also been reported to enhance the production of phytoalexin in suspension culture cells and also in transformed root cultures of *Hyoscyamus muticus*.[81] It was also effective in inducing taxoid production by *Taxus* cells.[82,83] Salicylic acid also affected the synthesis of rosmarinic acid and related enzyme activities in suspension cultures of *Salvia miltiorrhiza*.[84]

S-Methyl benzo-1,2,3-thiadiazole-7-carboxylate (BTH), the first commercial SAR activator for plants, was able to induce the same set of defense responses as salicylic acid in signal transduction.[85,86] Afterwards, it was proved that BTH could also induce secondary metabolism in plant cell cultures besides their reported effect on crop protection. Pretreatment with BTH can augment the sensitivity for low-dose elicitation of coumarin phytoalexin secretion by cultured parsley (*Petroselinum crispum* L.) cells.[87] BTH can be used to enhance rosmarinic acid production in a suspension culture of *Agastache rugosa*.[88]

Typical compounds 2,6-dichloro-isonicotinic acid (INA) and its methyl ester were also able to induce SAR in plants. It was reported that after *Peronospora parasitia* invaded *Arabidopsis* treated with INA, its symptoms were similar to an allergic reaction.[89,90] However, the cytotoxicity of INA was relatively strong; even within the range of its effective concentration there was still large toxicity.

6.2.4 Synthetic Organic Elicitors

In addition to various fungi and cell wall fragments, natural small organic molecules, such as JA and its esters, salicylic acid and some analogs, are also widely used. Based on the combination of synthetic chemistry and biotechnology, Qian *et al.* designed and synthesized a series of novel chemical organic elicitors according to the nature of biological macromolecules in plant cells and the mechanisms of action of elicitors. Highly effective elicitors were confirmed in plant cell culture, including synthetic JA and MJA derivatives,[91–95] synthetic SA derivatives,[96] synthetic INA derivatives,[97] and synthetic benzothiadiazole (BTH) derivatives.[98] These works provide a good precedent of using chemical methods to solve biological problems.

6.2.4.1 Synthetic JA and MJA derivatives

Some studies have shown that a little diversification on the structure of jasmonates could make a great difference to the induction effect. In 1996, Yukimune *et al.* reported that the carboxyl group of the C-1 position and the carbonyl group of the C-6 position in MJA were important for promoting the accumulation of taxanes in suspension cell culture.[36] They found that the induced production of paclitaxel was down sharply when carbonyl group on the MJA C-6 position was reduced to hydroxyl. If the C-1 position carboxyl group was lost, its inducing activity would also disappear in the production of paclitaxel. In 2000, Yukimune *et al.* also found that MJA configuration had a significant effect on the synthesis of *Taxus* media cell growth and taxane metabolites.[99] Miersch *et al.*[100] reported that JA branched at the C-7 position was very important for the induction of gene expression in barley leaf after investigating 66 JA structural analogs. In *Taxus chinensis* cell suspension cultures, the induction of taxane production using HMJA (6-hydroxy) was about 25% lower compared to MJA,[101] while the esterification of JA at the C-1 position had a significant impact on taxane induction in *T. chinensis* cell suspension cultures.[102,103] Qian *et al.* synthesized a series of MJA derivatives and made a very significant induction effect in plant cell culture.[91–94] A series of successful synthetic MJA derivatives are listed in Figure 6.5.

6.2.4.2 Synthetic Benzothiadiazole (BTH) Derivatives

It was noticed that *S*-methyl benzo-1,2,3-thiadiazole-7-carboxylate (BTH), the first commercial SAR activator for plants, could induce a similar defense response to salicylic acid in signal transduction. Therefore, it was interesting to investigate whether or not BTH derivatives could also induce secondary

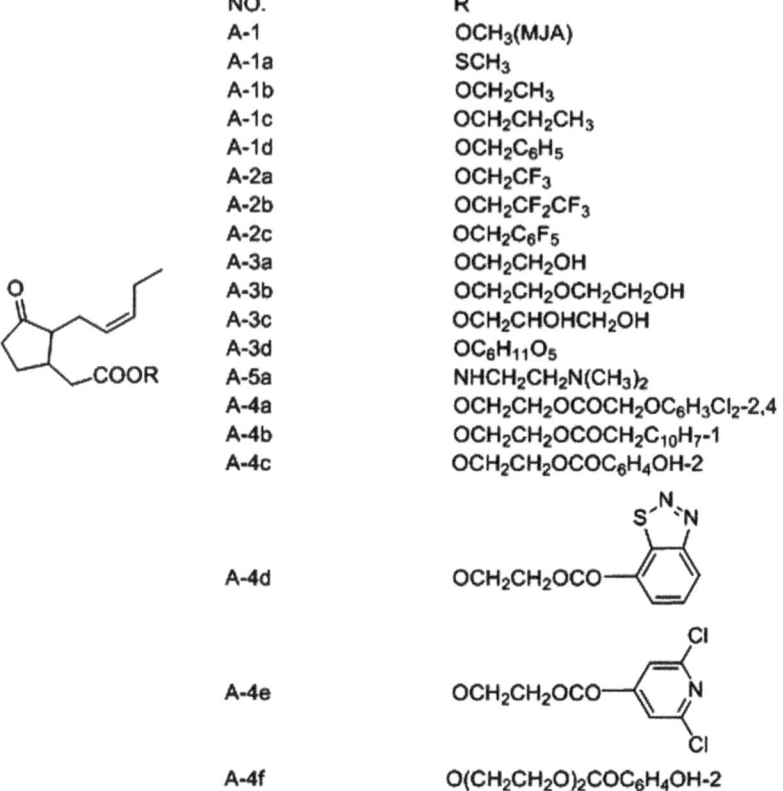

NO.	R
A-1	OCH$_3$(MJA)
A-1a	SCH$_3$
A-1b	OCH$_2$CH$_3$
A-1c	OCH$_2$CH$_2$CH$_3$
A-1d	OCH$_2$C$_6$H$_5$
A-2a	OCH$_2$CF$_3$
A-2b	OCH$_2$CF$_2$CF$_3$
A-2c	OCH$_2$C$_6$F$_5$
A-3a	OCH$_2$CH$_2$OH
A-3b	OCH$_2$CH$_2$OCH$_2$CH$_2$OH
A-3c	OCH$_2$CHOHCH$_2$OH
A-3d	OC$_6$H$_{11}$O$_5$
A-5a	NHCH$_2$CH$_2$N(CH$_3$)$_2$
A-4a	OCH$_2$CH$_2$OCOCH$_2$OC$_6$H$_3$Cl$_2$-2,4
A-4b	OCH$_2$CH$_2$OCOCH$_2$C$_{10}$H$_7$-1
A-4c	OCH$_2$CH$_2$OCOC$_6$H$_4$OH-2
A-4d	OCH$_2$CH$_2$OCO—
A-4e	OCH$_2$CH$_2$OCO—
A-4f	O(CH$_2$CH$_2$O)$_2$COC$_6$H$_4$OH-2

Figure 6.5 Synthetic MJA derivatives.

metabolism in plant cell cultures. Therefore, a series of novel BTH derivatives were designed and synthesized by our group. The synthetic structures are shown in Figure 6.6.

6.3 Application of Chemical Bioengineering in Plant Cell Culture

6.3.1 Taxane Productions from *Taxus chinensis* Cell Cultures

Taxanes are diterpene compounds containing a taxane skeleton. Paclitaxel, the first compound identified with a taxane ring, was isolated from the bark of the pacific yew, *Taxus brevifolia* in 1971.[102] Subsequently, paclitaxel and taxoid derivatives have been reported from the foliage and bark of several other species of *Taxus*, like *T. wallichinan*, *T. baccata*, *T. canadensis*, *T. cuspidata*, and *T. Yunnanensis*.[103–107] In addition to the plant resources, some endophytic fungi, such as *Tubercularia* sp., *Sporormia minima* and *Seimatoantlerium tepuiense*, have also been reported to produce paclitaxel and other taxoids.[108–110] Paclitaxel is effective for the treatment of leukemia and cancers. It has been reported that paclitaxel is capable of curing approximately 30%, 50%, and 20% of ovarian, breast, and lung cancer patients, respectively.[111,112]

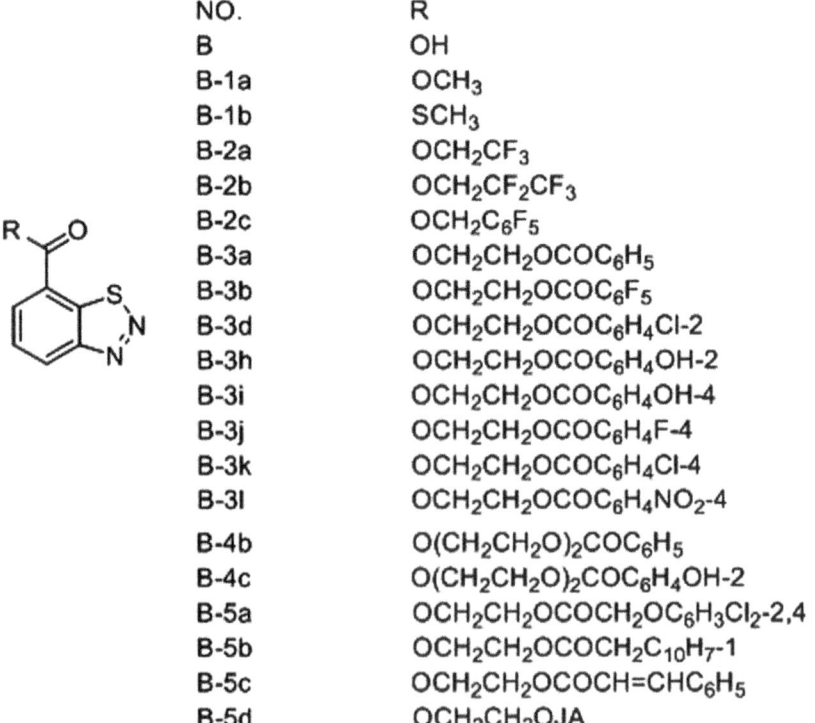

NO.	R
B	OH
B-1a	OCH_3
B-1b	SCH_3
B-2a	OCH_2CF_3
B-2b	$OCH_2CF_2CF_3$
B-2c	$OCH_2C_6F_5$
B-3a	$OCH_2CH_2OCOC_6H_5$
B-3b	$OCH_2CH_2OCOC_6F_5$
B-3d	$OCH_2CH_2OCOC_6H_4Cl\text{-}2$
B-3h	$OCH_2CH_2OCOC_6H_4OH\text{-}2$
B-3i	$OCH_2CH_2OCOC_6H_4OH\text{-}4$
B-3j	$OCH_2CH_2OCOC_6H_4F\text{-}4$
B-3k	$OCH_2CH_2OCOC_6H_4Cl\text{-}4$
B-3l	$OCH_2CH_2OCOC_6H_4NO_2\text{-}4$
B-4b	$O(CH_2CH_2O)_2COC_6H_5$
B-4c	$O(CH_2CH_2O)_2COC_6H_4OH\text{-}2$
B-5a	$OCH_2CH_2OCOCH_2OC_6H_3Cl_2\text{-}2,4$
B-5b	$OCH_2CH_2OCOCH_2C_{10}H_7\text{-}1$
B-5c	$OCH_2CH_2OCOCH=CHC_6H_5$
B-5d	OCH_2CH_2OJA

Figure 6.6   A series of synthetic BTH derivatives.

It is now known that there are more than 380 kinds of natural taxane in *Taxus*. The biosynthesis of taxanes is generally believed to occur as follows. Farnesyl diphosphate (FPP) and isopentenyl pyrophosphate (IPP) are first catalyzed by GGPP synthase to produce geranylgeranyl-pyrophosphate (GGPP), and then taxa-4(5),11(12)-diene is produced under the action of taxa-diene synthase.[113] Next, taxanes are produced by the hydroxylase oxidation of the taxa-4(20),11(12)-dien-5α-ol diterpene parent ring.[114–116] The biosynthetic pathway for taxanes is illustrated in Figure 6.7.

To obtain paclitaxel, total chemical synthesis, semi-synthesis, and extraction methods have been employed by excellent synthetic chemists. Total synthesis, however, has not been practically applied in the industry because of the too complex structure. A semi-synthetic method employing precursors such as 10-deacetylbaccatin III is now used in industry to obtain paclitaxel and has revealed some drawbacks since 10-deacetylbaccatin III is extracted from a large number of yew trees (Figure 6.8), and mass deforestation has led to serious environmental destruction.

In view of the above conditions, cell culture has been proposed as a promising alternative method to obtain paclitaxel. Cell culture processes can be adapted to produce paclitaxel as well as the taxane precursor baccatin. However, the yield of paclitaxel or baccatin obtained using cell cultures is too low to be sufficient for industrial application. Therefore, there is an urgent need to develop effective means to increase paclitaxel production on an industrial scale.

In order to improve the yield of paclitaxel in cell cultures, many methods have been tried in which elicitation was very significant. According to the work of Yukimune *et al.*, adding 100 μM MJA increased paclitaxel production 5.1-fold, reaching 115.2 mg L^{-1}.[36] Subsequent results have proved that the configuration of MJA could affect paclitaxel and baccatin III production in *Taxus* cells. So, a series of synthetic MJA derivatives were synthesized by Qian *et al.*[92] A suspension culture of *Taxus chinensis* was used as a model plant cell system to evaluate novel synthetic jasmonates as elicitors for stimulating

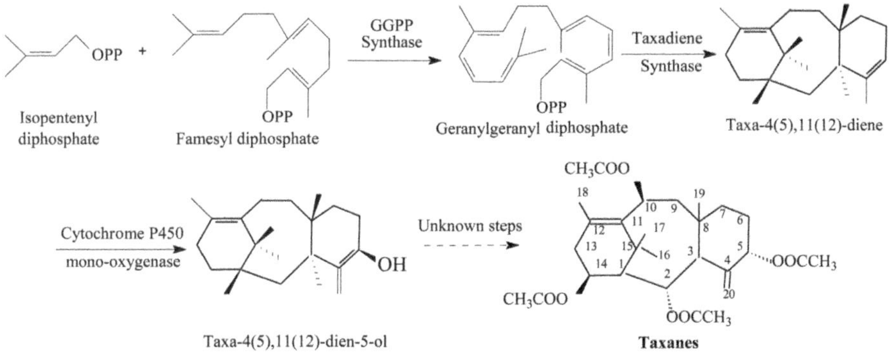

Figure 6.7 The biosynthetic pathway for taxanes.

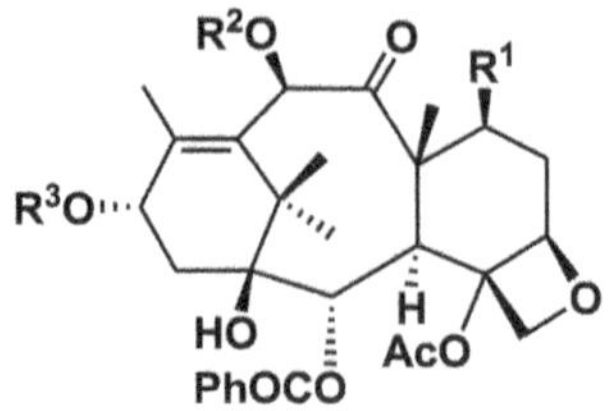

Paclitaxel	R^1=OH; R^2=Ac; R^3=
10-Deacetyltaxol	R^1=OH; R^2=H; R^3=
Baccatin III	R^1=OH; R^2=Ac; R^3=H
10-Deacetybaccatin III	R^1=OH; R^2=H; R^3=H

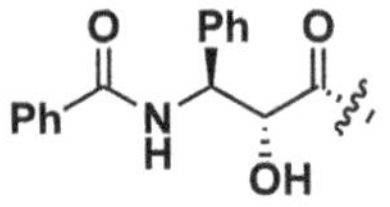

Figure 6.8   Chemical structures of taxoids.

the biosynthesis of secondary metabolites. Accumulations of taxuyunnanine C (Tc) were significantly increased with the addition of newly synthesized 2-hydroxyethyl jasmonate (HEJA) and trifluoroethyl jasmonate (TFEJA). With the addition of 100 μM HEJA (A-3c) or TFEJA (A-4d) on day 7, high Tc production was induced and about 47.2 mg g^{-1} or 39.7 mg g^{-1} (at day 21) of Tc content was obtained by the addition of 100 μM methyl jasmonate (MJA), while the Tc content was 13.7 mg g^{-1} and 29.2 mg g^{-1} for the control. The results are shown in Table 6.5. The new compounds HEJA and TFEJA were generally considered as the best elicitors for taxoid biosynthesis, which suggested that the novel jasmonate analogues may have great potential applications in other cell culture systems for efficient elicitation of plant secondary metabolites.

Unnatural elicitors in the form of BTH derivatives and INA derivatives were also successfully applied in *Taxus* cells (Tables 6.6 and 6.7). The results showed that the biosynthesis of some secondary metabolites is regulated by some small compounds. The chemical bioengineering strategy lays the foundation not only for the industrial application of paclitaxel, but also for other bioprocess systems.

6.3.2   Ginsenosides from *Panax notoginseng* Cell Cultures

Ginseng is a famous and precious herb that has been used as a healing drug and health tonic in China and some other oriental countries, including Korea and Japan, since ancient times. Ginseng saponins are important active ingredients from the ginseng genus, of which dammarenediol can generate ginsenosides such as Rg1 and Rb1 through a series of process modifications, such as oxidation, substitution and glycosylation.

Table 6.5   Measurement of inducing activity of MJA derivatives in *Taxus* cells.

Compound no.	Cell dry weight (g DW L^{-1})	Content of Tc mg (g DW)$^{-1}$	Maximum Tc productivity (mg L^{-1})
Control	16.4 ± 0.3	13.7 ± 1.4	154 ± 12
A-1	15.8 ± 0.1	29.2 ± 0.6	386 ± 10
A-1a	14.3 ± 0.6	31.5 ± 0.5	379 ± 16
A-1b	15.2 ± 0.3	28.3 ± 0.4	369 ± 9
A-1c	15.3 ± 0.0	27.9 ± 1.6	357 ± 7
A-1d	14.9 ± 0.4	26.7 ± 0.8	347 ± 11
A-2a	15.1 ± 0.3	34.94 ± 0.1	440.3 ± 5.9
A-2b	15.7 ± 0.3	38.22 ± 0.9	480.5 ± 0.6
A-2c	15.9 ± 0.0	34.6 ± 0.9	425 ± 6
A-3a	16.2 ± 0.4	38.1 ± 0.9	464 ± 20
A-3b	16.1 ± 0.4	31.4 ± 1.5	417 ± 2
A-3c	15.4 ± 0.4	47.2 ± 0.5	550 ± 13
A-3d	15.6 ± 0.4	40.5 ± 0.0	492 ± 6
A-4a	13.3 ± 0.1	29.5 ± 1.5	308 ± 9
A-4b	14.1 ± 0.5	28.9 ± 1.2	312 ± 6
A-4c	16.4 ± 0.1	31.6 ± 1.8	375 ± 16
A-4d	13.9 ± 0.2	39.6 ± 0.8	427.4 ± 15.4
A-4e	11.7 ± 0.5	27.7 ± 1.0	322.2 ± 3.8
A-4f	16.0 ± 0.1	36.8 ± 0.5	424 ± 17
A-5a	17.5 ± 0.6	21.0 ± 1.1	275 ± 4

Table 6.6   Measurement of inducing activity of BTH derivatives in *Taxus* cells.

Compound No.	Content of Tc mg (g DW)$^{-1}$	Maximum Tc productivity (mg L^{-1})
Ctrl.	11.7 ± 0.5	159 ± 4
B-1	16.2 ± 0.6	243.9 ± 9
B-1a	17.3 ± 0.7	274.9 ± 4
B-1b	16.8 ± 0.4	259.4 ± 9
B-2a	25.2 ± 0.7	299.6 ± 12
B-2b	16.2 ± 0.3	216.1 ± 2
B-2c	15.0 ± 0.4	199 ± 5
B-3a	14.7 ± 1.2	180.5 ± 3
B-3b	16.3 ± 0.3	213.7 ± 7
B-3d	22.9 ± 0.8	303.3 ± 11
B-3h	39.4 ± 3.3	534.8 ± 21
B-3i	15.6 ± 1.3	223.4 ± 14
B-3j	14.6 ± 0.5	197.4 ± 9
B-3k	15.1 ± 0.5	209.8 ± 11
B-3l	12.9 ± 0.4	187.2 ± 7
B-4b	14.3 ± 0.6	194.0 ± 9
B-4c	17.5 ± 0.7	199.9 ± 4
B-5a	15.5 ± 0.2	194.0 ± 3
B-5b	15.9 ± 0.2	193.2 ± 6
B-5c	14.5 ± 0.3	188.4 ± 8
B-5d	32.9 ± 0.8	457.5 ± 12
MJA	28.4 ± 0.1	390.6 ± 7
SA	17.3 ± 0.6	269.6 ± 20

Table 6.7 Effects of INA derivatives on cell growth, content of Tc, and yield of Tc.

Compound No.	Cell dry weight (g DW L^{-1})	Content of Tc mg (g DW)$^{-1}$	Maximum Tc productivity (mg L^{-1})
Control	15.1 ± 0.3	13.7 ± 1.4	154 ± 12
Isonicotinic acid	12.9 ± 0.4	13.3 ± 2.2	151 ± 5
Nicotinic acid	12.1 ± 0.7	18.9 ± 0.2	207 ± 4
INA	11.9 ± 0.3	17.1 ± 0.9	200 ± 1
C-1	13.8 ± 0.6	13.9 ± 1.1	153 ± 6
C-2	12.0 ± 0.2	19.7 ± 0.6	211 ± 2
C-3	13.1 ± 0.5	21.6 ± 2.0	257 ± 3
C-4	12.7 ± 0.4	14.2 ± 1.0	178 ± 2
C-5	12.4 ± 0.1	18.2 ± 0.8	209 ± 3

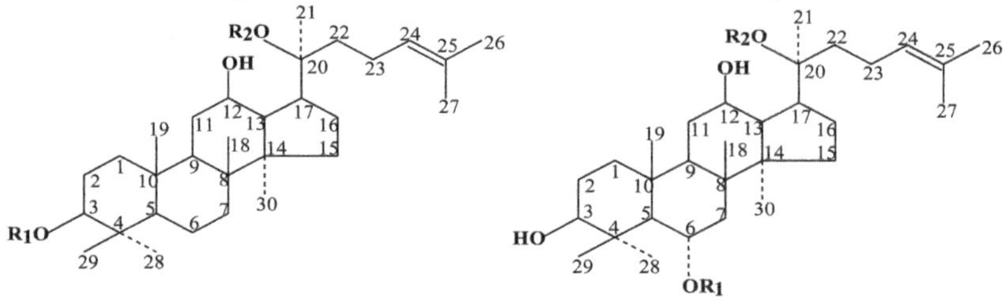

Figure 6.9 Chemical structures of protopanaxadiol and protopanaxatriol.

In *Panax ginseng*, ginsenoside is one of the main active ingredients, which are mainly composed of tetracyclic triterpenoid ginsenoside, in which the glycoside unit originates from protopanaxadiol or protopanaxatriol. The chemical structures are shown in Figure 6.9. The differences between the ginsenoside structures are mainly on the sugar residue. Ginsenosides Rg1 and Rb1 are the main ginsenosides in *Panax ginseng*.

Zhong *et al.* have constructed a more stable and productive cell line of *Panax notoginseng* and successfully applied a modified MS medium to achieve high cell density cultivation of *P. notoginseng* cells.[117–119] In *P. notoginseng*, synthetic elicitors can also induce the biosynthesis of ginsenosides. As can be seen from the data in Table 6.8, the impact of the addition of MJA derivatives on ginsenoside content and yield in *P. notoginseng* are evident. The compound A-3a had the best effect leading to a nearly 5-fold increase from the control (0.65 mg/100 mg DW) to 3.65 mg/100 mg DW after the induction. The four ginsenosides detected in protopanaxadiol ginsenoside Rb1 (protopanadiol) changed significantly before and after the induction, as shown in Table 6.8. The relationship between the elicitor structure and the Rb1 content are shown in Table 6.9 and are discussed further in the following paragraph.

Several compounds containing hydroxyl groups displayed better inducing activity in ginsenosides in *P. notoginseng*, which was similar to the results in

Table 6.8 Effects of jasmonate derivatives on ginsenoside content in *P. notoginseng* cell cultures.

Elicitors	DW (g L^{-1}) (day 13)	Ginsenoside content (mg/100 mg DW)				
		Rg$_1$	Re	Rb$_1$	Rd	Total[a]
Control	8.87 ± 0.25	0.20 ± 0.01	0.24 ± 0.02	0.21 ± 0.02	0	0.65 ± 0.02
MJA	7.94 ± 0.46	0.36 ± 0.02	0.44 ± 0.01	1.21 ± 0.04	0.07 ± 0.02	2.07 ± 0.04
A-1a	7.91 ± 0.25	0.39 ± 0.02	0.48 ± 0.02	1.21 ± 0.03	0.10 ± 0.03	2.18 ± 0.03
A-1c	8.47 ± 0.25	0.42 ± 0.02	0.48 ± 0.03	1.17 ± 0.02	0.09 ± 0.03	2.17 ± 0.05
A-2a	7.88 ± 0.17	0.30 ± 0.03	0.36 ± 0.02	0.89 ± 0.07	0.06 ± 0.02	1.62 ± 0.09
A-2b	7.59 ± 0.32	0.45 ± 0.01	0.53 ± 0.01	1.44 ± 0.05	0.13 ± 0.03	2.56 ± 0.03
A-2c	8.19 ± 0.27	0.29 ± 0.01	0.39 ± 0.02	0.87 ± 0.04	0.06 ± 0.01	1.60 ± 0.06
A-3a	8.15 ± 0.62	0.62 ± 0.02	0.66 ± 0.02	2.20 ± 0.06	0.18 ± 0.04	3.65 ± 0.08
A-3b	8.29 ± 0.38	0.56 ± 0.02	0.60 ± 0.02	1.69 ± 0.04	0.11 ± 0.03	2.96 ± 0.09
A-3c	8.87 ± 0.51	0.40 ± 0.01	0.52 ± 0.01	1.21 ± 0.07	0.06 ± 0.02	2.19 ± 0.11
A-3d	8.55 ± 0.17	0.33 ± 0.03	0.40 ± 0.02	0.93 ± 0.03	0.06 ± 0.02	1.73 ± 0.02
A-4a	6.49 ± 0.11	0.52 ± 0.02	0.53 ± 0.01	1.65 ± 0.05	0.13 ± 0.04	2.83 ± 0.10
A-4c	11.01 ± 0.21	0.09 ± 0.01	0.07 ± 0.01	0.09 ± 0.02	0	0.35 ± 0.02
A-4f	8.62 ± 0.65	0.04 ± 0.00	0.08 ± 0.01	0.05 ± 0.00	0	0.17 ± 0.01
A-5a	7.25 ± 0.31	0.24 ± 0.01	0.32 ± 0.02	0.44 ± 0.02	0.04 ± 0.01	1.04 ± 0.02

[a]Total = Rg$_1$ + Re + Rb$_1$ + Rd.

Table 6.9 Effects of MJA derivatives containing hydroxyl on the production of protopanaxadiol Rb1.

Compounds	Rb$_1$ content [mg (g DW)$^{-1}$]	Number of OH	AlogP98
MJA	1.21 ± 0.04	0	2.22
A-3a	2.20 ± 0.06	1	1.68
A-3b	1.69 ± 0.04	1	1.55
A-3c	1.21 ± 0.07	2	1.17
A-3d	0.93 ± 0.03	4	0.06

T. chinensis, but a different trend was observed in Table 6.9. A-3a and A-3b were the best elicitors, inducing Rb1 content to increase by 80% and 40%, respectively, under the same conditions compared to MJA. Two compounds containing one free hydroxyl group have approximately the same lipophilicity value. When the number of hydroxyls or the AlogP98 value moves away from 1.6, the induction activities of those synthetic compounds decreased ginsenoside production in *P. notoginseng*. The best active compound A-3c in *T. chinensis* cell cultures showed only some similarities with MJA in *P. notoginseng*. The results showed that there are great differences in the responses of different plant cells to diverse elicitors.

Data in our experiment (unpublished) showed that BTH derivatives exhibited inhibitory effects on both ginsenosides content and growth of *P. notoginseng* cells, which demonstrated that the elicitation effects differ from one species to another and from one metabolic pathway to another in the plant cell cultures (Table 6.10).

Table 6.10 Effects of BTH derivatives on the growth and ginsenoside content of
P. notoginseng cells.

Elicitors	DW (g L^{-1}) (day 13)	Ginsenoside content (mg/100 mg DW)				
		Rg$_1$	Re	Rb$_1$	Rd	Total
Control	8.87 ± 0.25	0.20 ± 0.01	0.24 ± 0.02	0.21 ± 0.02	0	0.65 ± 0.02
B-1a	9.72 ± 0.07	0.06	0.07	0.04	0	0.17
B-2a	9.11 ± 0.29	0.04	0.11	0.05	0	0.19
B-3h	9.24 ± 0.18	0.10	0.13	0.08	0	0.31
B-6a	10.52 ± 0.71	0.04	0.08	0.03	0	0.16
B-2b	9.58 ± 0.77	0.04	0.09	0.04	0	0.16
B-3d	9.61 ± 0.17	0.09	0.14	0.08	0	0.31
B-1b	9.03 ± 0.53	0.04	0.12	0.04	0	0.21
B-3i	10.09 ± 0.28	0.07	0.10	0.06	0	0.23
B-3k	10.26 ± 0.11	0.09	0.14	0.07	0	0.30
B-3j	9.55 ± 0.23	0.09	0.13	0.07	0	0.28

6.4 Perspectives

It is expected that the biosynthetic capacity of plants could be exploited *in
vitro* using plant cells and cell tissue systems, analogous to microbial cells in
fermentation processes. An important requirement for the improvement of
secondary metabolite synthesis is an understanding of the metabolic pathways and the enzymology of the biosynthesis of particular products. The
knowledge of plant metabolic pathways is still very limited. It needs more
in-depth study by biologists, which will help to make the chemical bioengineering more practical.

In-depth studies of pathways in whole plants are often difficult because
the biosynthetic activities may only be expressed in particular cell types
within a specific plant organ or at a certain time of year or season. Cell
cultures have a higher rate of metabolism than intact differentiated plants
because the initiation of cell growth in culture can reduce the biosynthetic
cycle. With the rise of synthetic biology technology, many plant secondary metabolites or the premises of plant secondary metabolites can be
expressed in a microorganism,[120,121] which will give chemical bioengineering a new platform to promote the application of synthetic elicitors in plant
cell culture.

References

1. W. Kotte, *Beitr. Allg. Bot.*, 1922, **2**, 413–434.
2. W. J. Robbins, *Bot. Gaz.*, 1922, **73**, 376–390.
3. W. Tulecke, *Bull. Torrey Bot. Club*, 1961, **88**, 350–360.
4. R. B. Buchanan, W. Gruissem and R. L. Jones, *Am. Soc. Plant Physiol.*,
    2000, 930–988.
5. R. Verpoorte, A. Contin and J. Memelink, *Phytochem. Rev.*, 2002, **1**,
    13–25.

6. S. M. Rates, *Toxicon*, 2001, **39**, 603–613.
7. J. B. Harborne, N. J. Walton and D. E. Brown, *Perspectives on plant secondary products*, Imperial College Press, London, 1999, pp. 1–25.
8. T. C. Zhou, W. W. Zhou, W. Hu and J. J. Zhong, *Encyclopedia of Industrial Biotechnology: Bioprocess, Bioseparation, and Cell Technology*, John Wiley & Sons, Inc., Hoboken, 2010, pp. 1–27.
9. R. S. Ramachandra and G. A. Ravishankar, *Biotechnol. Adv.*, 2002, **20**, 101–153.
10. I. Smetanska, *Adv. Biochem. Eng./Biotechnol.*, 2008, **111**, 187–228.
11. G. Z. Zheng and Z. R. Yang, *The Biology of Noteginseng and its application*, Science Press, 1994, pp. 1–6.
12. X. Zhou and J. J. Zhong, *Encyclopedia of Industrial Biotechnology: Bioprocess, Bioseparation, and Cell Technology*, 2010, pp. 1–29.
13. Y. Fujita and M. Tabata, *Plant Tissue and Cells Culture*, ed. Alan R. Liss, New York, 1987, pp. 169–185.
14. S. R. Rao and G. A. Ravishankar, *Biotechnol. Adv.*, 2002, **20**, 101–153.
15. B. Ulbrich, W. Wiesner and H. Arens, in *Primary and Secondary Metabolism of Plant Cell Cultures*, ed. K. H. Neumann, *et al.*, Springer Verlag, 1985, pp. 293–303.
16. M. A. Baque, S. H. Moh, E. J. Lee, J. J. Zhong and K. Y. Paek, *Biotechnol. Adv.*, 2012, **30**, 1255–1267.
17. G. Sembdner and B. Parthier, *Annu. Rev. Plant Physiol. Plant Mol. Biol.*, 1993, **44**, 569–589.
18. H. Dornenburg and D. Knor, *Enzyme Microb. Technol.*, 1995, **17**, 674–684.
19. J. J. Zhong, *Adv. Biochem. Eng./Biotechnol.*, 2001, **72**, 1–26.
20. J. Wu and J. J. Zhong, *J. Biotechnol.*, 1999, **68**, 89–99.
21. A. G. Namdeo, S. Patil and D. P. Fulzele, *Biotechnol. Prog.*, 2002, **18**, 159–162.
22. K. Y. Paek, H. N. Murthy, E. J. Hahn and J. J. Zhong, *Adv. Biochem. Eng./Biotechnol.*, 2009, **113**, 151–176.
23. M. Vanishree, C. Y. Lee, S. F. Lo, S. M. Nalawade, C. Y. Lin and H. S. Tsay, *Bot. Bull. Acad. Sin.*, 2004, **45**, 1–22.
24. J. J. Zhong, in *Comprehensive Biotechnology*, ed. M. Moo-Young, 2nd edn, 2011, vol. 3, pp. 299–308.
25. M. I. Georgiev, R. Eibl and J. J. Zhong, *Appl. Microbiol. Biotechnol.*, 2013, **97**, 3787–3800.
26. F. Dicosmo and M. Misawa, *Biotechnol. Adv.*, 1995, **13**, 425–435.
27. M. H. Zenk, *Int Ass for Plant Tissue Culture*, Calgary, 1978, pp. 1–14.
28. L. G. Romagnoli and D. Knorr, *Food Biotechnol.*, 1988, **2**, 93–104.
29. J. Zhao, L. Davis and R. Verpoorte, *Biotechnol. Adv.*, 2005, **23**, 283–333.
30. R. Radman, T. Saez, C. Bucke and T. Keshavarz, *Biotechnol. Appl. Biochem.*, 2003, **37**, 91–102.
31. K. M. Oksman-Caldentey and R. Hiltunen, *Field Crop. Res.*, 1996, **45**, 57–69.
32. F. Dicosmo and M. Misawa, *Trends Biotechnol.*, 1985, **3**, 318–322.
33. J. Ebel and E. G. Cosio, *Int. Rev. Cytol.*, 1994, **148**, 1–36.
34. P. Komaraiah, P. B. K. Kishor, M. Carlsson, K. E. Magnusson and C. F. Mandenius, *Plant Sci.*, 2005, **168**, 1337–1344.

35. C. W. T. Lee-Parsons, S. Erturk and J. Tengtrakool, *Biotechnol. Lett.*, 2004, **26**, 1595–1599.
36. Y. Yukimune, H. Tabata, Y. Higashi and Y. Hara, *Nat. Biotechnol.*, 1996, **14**, 1129–1132.
37. M. Yoshikawa, N. T. Keen and M. C. Wang, *Plant Physiol.*, 1983, **73**, 497–506.
38. Y. Ito, H. Kaku and N. Shibuya, *Plant J.*, 1997, **12**, 347–356.
39. S. Parchmann, H. Gundlach and M. J. Mueller, *Plant Physiol.*, 1997, **115**, 1057–1064.
40. D. Nennstiel, D. Scheel and T. Nqrnberger, *FEBS Lett.*, 1998, **431**, 405–410.
41. J. Lee, D. F. Klessig and T. Nurnberger, *Plant Cell*, 2001, **13**, 1079–1093.
42. C. Ji, D. Boyd, Y. Slaymaker, Y. Okinaka, S. L. Takeuchi, E. J. Herman and N. Keen, *Proc. Natl. Acad. Sci. U. S. A.*, 1998, **95**, 3306–3311.
43. S. Bourque, M. N. Binet, M. Ponchet, A. Pugin and A. Lebrun-Garcia, *J. Biol. Chem.*, 1999, **274**, 34699–34705.
44. J. Fliegmann, G. Schuler, W. Boland, J. Ebel and A. Mithofer, *Biol. Chem.*, 2003, **384**, 437–446.
45. C. Wasternack and D. X. Xie, *Plant Signaling Behav.*, 2010, **5**, 337–340.
46. J. Browse, *Annu. Rev. Plant Biol.*, 2009, **60**, 183–205.
47. G. Suzuki, H. Ohta, T. Kato, T. Igarashi, F. Sakaic, D. Shibatab, A. Takanoa, T. Masudaa, Y. Shioia and K. I. Takamiya, *FEBS Lett.*, 1996, **383**, 83–86.
48. F. X. Hu, J. H. Huang, Y. F. Xu, X. H. Qian and J. J. Zhong, *Biotechnol. Bioeng.*, 2006, **94**, 1064–1071.
49. F. X. Hu and J. J. Zhong, *Process Biochem.*, 2008, **43**, 113–118.
50. A. William, G. John and J. Hendel, *J. Chromatogr.*, 1996, **775**, 11–17.
51. H. Mizukami, Y. Tavira and B. E. Ellis, *Plant Cell Rep.*, 1993, **12**, 706–709.
52. F. X. Hu and J. J. Zhong, *J. Biosci. Bioeng.*, 2007, **104**, 513–516.
53. H. Grundlach, M. J. Müller, T. M. Kutchan and M. H. Zenk, *Proc. Natl. Acad. Sci. U. S. A.*, 1992, **89**, 2389–2393.
54. C. B. Do and F. Cormier, *Plant Cell Rep.*, 1990, **9**, 143–146.
55. Y. H. Zhang, J. J. Zhong and J. T. Yu, *Biotechnol. Lett.*, 1995, **17**, 1347–1350.
56. R. Rakwal, S. Tamogami and O. Kodama, *Biosci., Biotechnol., Biochem.*, 1996, **60**, 1046–1048.
57. S. I. Kim, H. K. Choi, J. H. Kim, H. S. Lee and S. S. Hong, *Enzyme Microb. Technol.*, 2001, **28**, 202–209.
58. C. Veersham, in *Elicitation: Med Plant Biotechnol*, CBS Publisher, India, 2004, pp. 270–293.
59. M. B. Ali, E. J. Hahn and K. Y. Paek, *Plant Cell Rep.*, 2006, **25**, 1122–1132.
60. H. K. Choi, T. L. Adams, R. W. Stahlhut, S. Kim, J. H. Yun, B. K. Song, J. H. Kim, J. S. Song, S. S. Hong, S. H. Lee and H. J. Choi, *US Pat.*, 6248572, 2001.
61. H. G. Wu, Z. Ma, Y. D. Wang and Y. J. Yuan, *J. Chin. Rare Earth Soc.*, 2000, **18**, 360–362.
62. C. Huang and J. J. Zhong, *Process Biochem.*, 2013, **48**, 1227–1234.
63. Z. Sabine, N. Thorsten and F. Jean-Nalie, *Proc. Natl. Acad. Sci. U. S. A.*, 1997, **94**, 2751–2755.

64. E. Demole, E. Lederer and D. Mercier, *Helv. Chim. Acta*, 1962, **45**, 675–865.
65. D. C. Aldridge, S. Galt, D. Giles and W. B. Turner, *J. Chem. Soc. C*, 1971, 1623–1627.
66. J. Veda and J. Kato, *Plant Physiol.*, 1980, **66**, 246–249.
67. W. Dathe, H. Ronsch, A. Preiss, W. Schade, G. Sembdner and K. Schreiber, *Planta*, 1981, **153**, 530–535.
68. H. Yamane, H. Takagi, H. Abe, T. Yokota and N. Takahashi, *Plant Cell Physiol.*, 1981, **22**, 689–697.
69. A. J. Enyedi, N. YaIpani, P. Silverman and I. Raskin, *Cell*, 1992, **70**, 879–886.
70. B. A. Vick and D. C. Zimmerman, *Plant Physiol.*, 1984, **75**, 458–461.
71. R. E. B. Ketchum, D. M. Gibson, R. B. Croteau and M. L. Shuler, *Biotechnol. Bioeng.*, 1999, **62**, 97–105.
72. N. Mirjalili and J. C. Linden, *Biotechnol. Prog.*, 1996, **12**, 110–118.
73. H. Gundlach, M. J. Muller, M. J. Kutchan and M. H. Zenk, *Proc. Natl. Acad. Sci. U. S. A.*, 1993, **89**, 2389–2393.
74. R. J. Aerts, A. Schafer, M. Hesse, T. W. Baumann and A. Slusarenko, *Phytochemistry*, 1996, **42**, 417–422.
75. L. Saenz-Carbonell and V. M. Loyola-Vargas, *Appl. Biochem. Biotechnol.*, 1996, **61**, 321–331.
76. S. C. Roberts and M. L. Shuler, *Curr. Opin. Biotechnol.*, 1997, **8**, 154–159.
77. J. Durner, J. Shah and D. F. Klessig, *Trends Plant Sci.*, 1997, **2**, 1360–1385.
78. I. Raskin, *Plant Physiol.*, 1992, **99**, 799–803.
79. X. Zhou and J. J. Zhong, *Appl. Microbiol. Biotechnol.*, 2011, **90**, 1027–1036.
80. X. Zhou and J. J. Zhong, *Biotechnol. Bioeng.*, 2011, **108**, 216–221.
81. U. Mehmetoglu and W. R. Curtis, *Appl. Biochem. Biotechnol.*, 1997, **67**, 71–77.
82. Y. D. Wang, Y. J. Yuan and J. C. Wu, *Biochem. Eng. J.*, 2004, **19**, 259–265.
83. L. J. Yu, W. Z. Lan, W. M. Qin and H. B. Xu, *Process Biochem.*, 2001, **37**, 477–482.
84. M. Jiao, R. Cao, H. Chen, W. Hao and J. Dong, *Chin. J. Biotechnol.*, 2012, **28**, 320–328.
85. R. Schurter, W. Kunz and R. Nyfeler, *Eur. Pat.*, EP 313512, 1989.
86. J. S. Thaler, A. L. Fidaantsef, S. S. Duffey and R. M. Bostock, *J. Chem. Ecol.*, 1999, **25**, 1597–1608.
87. V. A. Katz, O. U. Thulke and U. Conrath, *Plant Physiol.*, 1998, **117**, 1333–1339.
88. H.-K. Kim, S.-R. Oh, H.-K. Lee and H. Huh, *Biotechnol. Lett.*, 2001, **23**, 55–60.
89. J. P. Métraux, G. L. Ahi, T. Staub, J. Speich, A. Steinemann and J. Ryals, *Adv. Mol. Genet. Plant-Microbe Interact.*, 1991, **1**, 432–439.
90. J. Dumer and D. F. Klessig, *Proc. Natl. Acad. Sci. U. S. A.*, 1995, **92**, 11312–11316.
91. Z. G. Qian, Z. J. Zhao, Y. F. Xu, X. H. Qian and J. J. Zhong, *Biotechnol. Bioeng.*, 2004, **86**, 809–816.

92. Z. G. Qian, Z. J. Zhao, Y. F. Xu, X. H. Qian and J. J. Zhong, *Biotechnol. Bioeng.*, 2004, **86**, 594–599.
93. Z. G. Qian, Z. J. Zhao, Y. F. Xu, X. H. Qian and J. J. Zhong, *Biotechnol. Bioeng.*, 2005, **90**, 516–521.
94. Z. G. Qian, Z. J. Zhao, Y. F. Xu, X. H. Qian and J. J. Zhong, *Appl. Microbiol. Biotechnol.*, 2005, **68**, 98–103.
95. Z. J. Zhao, Y. F. Xu, Z. G. Qian, W. H. Tian, X. H. Qian and J. J. Zhong, *Bioorg. Med. Chem. Lett.*, 2004, **14**, 4755–4758.
96. Z. G. Qian, Z. J. Zhao, Y. F. Xu, X. H. Qian and J. J. Zhong, *Biotechnol. Prog.*, 2006, **22**, 331–333.
97. Z. G. Qian, Z. J. Zhao, Y. F. Xu, X. H. Qian and J. J. Zhong, *Appl. Microbiol. Biotechnol.*, 2006, **71**, 164–167.
98. Y. F. Xu, Z. J. Zhao, Z. G. Qian, W. H. Tian, X. H. Qian and J. J. Zhong, *J. Agric. Food Chem.*, 2006, **54**, 8793–8798.
99. Y. Yukimune, Y. Hara, E. Nomura, H. Seto and S. Yoshida, *Phytochemistry*, 2000, **54**, 13–17.
100. O. Miersch, R. Kramell, B. Parthier and C. Wasternack, *Phytochemistry*, 1999, **50**, 353–361.
101. H. D. Dong and J. J. Zhong, *Biochem. Eng. J.*, 2001, **8**, 145–150.
102. E. Baloglu and D. G. I. Kingston, *J. Nat. Prod.*, 1999, **62**, 1448–1472.
103. D. G. I. Kingston, P. G. Jagtap, H. Yuan and L. Samala, *Prog. Chem. Org. Nat. Prod.*, 2002, **84**, 53–225.
104. M. Hezari, N. G. Lewis and R. Croteau, *Arch. Biochem. Biophys.*, 1995, **322**, 437–444.
105. J. Hefner, S. M. Rubenstein, R. E. B. Ketchum, D. M. Gibson, R. M. Willians and R. Croteau, *Chem. Biol.*, 1996, **3**, 479–489.
106. K. Walker and R. Croteau, *Phytochemistry*, 2001, **58**, 1–7.
107. S. Jennewein and R. Croteau, *Appl. Microbiol. Biotechnol.*, 2001, **57**, 13–19.
108. D. G. I. Kingston, *Trends Biotechnol.*, 1994, **12**, 222–227.
109. R. W. Miller, R. G. Powell, C. R. Smith, E. Arnold and J. Clardy, *J. Org. Chem.*, 1981, **46**, 1469–1474.
110. K. M. Witherup, S. A. Look, M. W. Stasko, T. J. Ghiorzi and G. M. Muschik, *J. Nat. Prod.*, 1990, **53**, 1249–1255.
111. A. G. Fett-Neto and F. DiCosmo, *Planta Med.*, 1992, **58**, 464–466.
112. H. N. ElSohly, E. D. Croom, W. J. Kopycki, A. S. Joshi, M. A. ElSohly and J. D. McChesney, *Phytochem. Anal.*, 1995, **6**, 149–156.
113. B. Singh, R. K. Gujral, R. P. Sood and H. Duddeck, *Planta Med.*, 1997, **63**, 191–192.
114. G. A. Strobel, E. Ford, J. Y. Li, J. Sears, R. S. Sidhu and W. M. Hess, *Appl. Microbiol.*, 1999, **22**, 426–433.
115. J. Wang, G. Li, H. Lu, Z. Zheng, Y. Huang and W. Su, *FEMS Microbiol. Lett.*, 2000, **193**, 249–253.
116. K. Shrestha, G. A. Strobel, S. Prakash and M. Gewali, *Planta Med.*, 2001, **67**, 374–376.
117. Y. H. Zhang and J. J. Zhong, *Enzyme Microb. Technol.*, 1997, **21**, 59–63.

118. W. W. Hu, H. Yao and J. J. Zhong, *Biotechnol. Prog.*, 2001, **17**, 838–846.
119. J. Han and J. J. Zhong, *Biotechnol. Lett.*, 2002, **24**, 1927–1930.
120. P. K. Ajikumar, W. H. Xiao, K. E. Tyo, Y. Wang, F. Simeon, E. Leonard, O. Mucha, T. H. Phon, T. Heng, B. Pfeifer and G. Stephanopoulos, *Science*, 2010, **330**, 70–74.
121. J. W. Lee, D. Na, J. M. Park, J. M. Lee, S. Choi and S. Y. Lee, *Nat. Chem. Biol.*, 2012, **8**, 536–546.

CHAPTER 7

Chemical Biotechnology for Plant Protection

ZHENJIANG ZHAO[a], YUFANG XU[a,b], AND XUHONG QIAN*[b]

[a]Shanghai Key Laboratory of New Drug Design, School of Pharmacy, East China University of Science and Technology, Shanghai 200237, China; [b]Shanghai Key Laboratory of Chemical Biology and State Key Laboratory of Bioreactor Engineering, School of Pharmacy, East China University of Science and Technology, Shanghai 200237, China
*E-mail: xhqian@ecust.edu.cn

7.1 Biotechnology in Agriculture

With the increase in the world's population, grain supply has become a serious problem. Contributing factors include fast population growth, low production output of grains and natural calamities. As the Food and Agriculture Organization of the United Nations suggested in 1991 (http://www.fao.org/), China faces a great challenge of providing enough food for more than 1.1 billion people with a limited amount of arable land. In 1995, Lester Brown predicted that China might face a grain shortage in his book titled "Who will feed China".[1] It is a great achievement that gross grain production in China has increased over the past ten years and a shortage of food has not been an immediate crisis. Nevertheless, food security is a long-term challenge to be dealt with, as was recognized by the China Central Economic Working Conference held in Beijing on October 13, 2013, in which national food security was still regarded as a prime matter.

RSC Green Chemistry No. 34
Chemical Biotechnology and Bioengineering
By Xuhong Qian, Zhenjiang Zhao, Yufang Xu, Jianhe Xu, Y.-H. Percival Zhang, Jingyan Zhang, Yangchun Yong, and Fengxian Hu
© X.-H. Qian, Z.-J. Zhao, Y.-F. Xu, J.-H. Xu, Y.-H. P. Zhang, J.-Y. Zhang, Y.-C. Yong and F.-X. Hu, 2015
Published by the Royal Society of Chemistry, www.rsc.org

In just fifty years from 1950 to 2000, the world population has grown from 2.5 billion to 6.1 billion. Currently, the world population grows by about 80 million per year, which demands extra grain output of at least about 40 million tons (0.5 kg cereal per day for one person) without consideration of the nutrient level. This increase, and the concomitant demand for grain, especially corn, wheat, rice and soybeans, will continue to place tremendous pressure on today's global agricultural production systems.[2]

To increase food production per capita, the following options could be applied separately or in combination: increasing the area of agricultural land, the use of agrichemicals, organic fertilizers, and biological controls, and improved soil and water management. Notably, new biotechnology has brought about genetically modified crops that are more productive and resistant to pests and diseases. Results are encouraging, but controversies persist.

7.1.1   Arable Land

To increase the area of land available for agriculture is not an easy task. In fact, agricultural land (hectares per inhabitant) is decreasing in most regions worldwide. For example, Latin America and the Caribbean witnessed a decrease from 0.40 ha per inh in 1990 to 0.32 ha per inh in 2010. North Africa and the Middle East also experienced a drop from 0.28 to 0.18 ha per inh in the same period. South Asia similarly suffered a decrease from 0.22 to 0.16 ha per inh. China takes fourth place for total arable land, behind Russia, America and India, but the per capita arable land is 0.106 ha per inh, which puts China in 126th position, less than half of the global average of 0.22 ha per inh. This is partly due to population growth but there is also a net loss of agricultural land due to erosion, reduction of fertility, salinization and desertification of soils. New land could be found only at the cost of sacrificing forest areas, many of which have been classified as ecological reserves and natural parks.[3]

7.1.2   Water Management

The total reserve of water on the earth is about 13.8×10^{11} tonnes, 97% of which is seawater. Within the remaining 3% of fresh water, 70% is distributed at the North Pole, South Pole or plateaus in the form of glaciers and icecaps, and the other 30% of fresh water is under the surface of the Earth as ground water and soil water. Water in lakes and swamps accounts for 0.29%, rivers 0.01% and the atmosphere covers 0.04%. Only 0.2% of the total water on the earth is available to mankind, including river water, fresh lake water and shallow ground water. Therefore, fresh water is a precious and limited resource.[4]

Today, water shortage has become a serious issue in Africa, the Middle East and Asia. Water for drinking and for agricultural irrigation is scarce. Water consumption is high in agriculture: production of 100 kg of wheat requires 50 000 L of water, and 100 kg of rice requires 200 000 L of water. Water resources available per capita are decreasing globally and for irrigation many countries are already using "fossil" waters pumped out from deep aquifers that will be exhausted in 20–30 years. Better management of water resources to increase water use efficiency is needed in many regions.

7.1.3   Agrichemicals: Fertilizers and Pesticides

Now, agrichemical use is probably the immediate answer to increasing grain production. Agrichemicals include two kinds of compounds: chemical fertilizers and pesticides. The use of fertilizers has tremendously increased worldwide food production since the 1960s. "How much of crop production is attributable to fertilizer input?" The answer to this question has been estimated to range from about 30 to 50% for major grain crops.[5] One-third of the increase in cereal production worldwide and half of the increase in India's grain production during the 1970s and 1980s have been attributed to increased fertilizer consumption.[6] Fertilizers (N, P, K and other microelements) can balance the element contents in soil in different regions to benefit plant growth, which has made a great contribution to increasing crop production. However, abuse of fertilizers, especially nitrate fertilizers, has caused serious contamination of aquifers, decreasing the quality of water.[7] The massive production of fertilizers, such as phosphate fertilizers from phosphorite and phosphoric acid, has led to problems of environmental pollution with heavy metals, such as cadmium and radionuclides of the uranium and thorium series.[8] So, increasing production using fertilizers is not sustainable.

Pesticides, including insecticides, fungicides, herbicides and rodenticides, are used to protect crops from pests, significantly reducing the losses to pests and improving the yields of crops such as corn, maize, vegetables, potatoes and cotton. An indisputable fact is that there has been a continuous increase in pesticide usage, both in number of chemicals and quantities, since the invention of pesticides. It was reported that pesticide consumption worldwide has increased from $850 million in 1960 to about $31 billion in 2005.[9]

Early pesticides from natural substances have been used for centuries. The famous Salvarsan, an organoarsenic compound, was invented in 1907. It induced the development of organic mercury compounds, which became extremely successful as seed disinfectants. When organic pesticides were introduced about sixty years ago, great expectations were raised by the invention of pesticides such as DDT in Figure 7.1.[10]

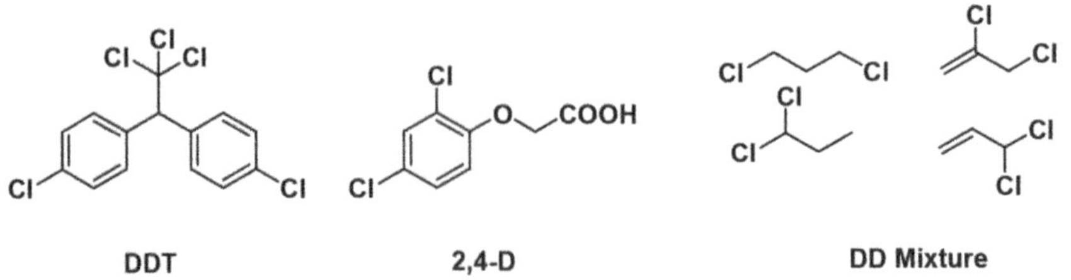

Figure 7.1   Chloro-containing pesticides or herbicides.

"A bright new star, a nova, named DDT had just burst brilliantly into the plant protection heavens. It was accompanied by some bright satellites; the dithiocarbamates for plant disease control. 2,4-D for weed control and DD for nematodes. Crop plants have never been so free of pests since agriculture was established. In the leaf hopper areas of America, potatoes have never been so green in September. The yields were doubled, often quadrupled."[11]

Let's review the story of DDT to understand the development of pesticides. 4,4-Dichloro diphenyl trichloromethane, not called DDT then, was discovered in 1873 by an Austrian graduate student Othmar Ziedler, who at the time was working in Adolf von Baeyer's laboratory; however, he never investigated its biological activity. It was a Swiss chemist, Paul Müller, who discovered that this compound had extraordinary insecticidal properties in September 1939. From then, DDT or Gesarol (named by Müller) has been applied not only as a kind of wonderful pesticide but also as an effective anti-malarial medicine during World War II with great success. In his Nobel awards acceptance speech in 1948, Paul Müller concluded that the ideal pesticide would have the following qualities: (1) great insect toxicity; (2) rapid onset of toxic action; (3) little or no mammalian or plant toxicity; (4) no irritant effect and no unpleasant odour; (5) the range of action should be as wide as possible, and cover as many arthropods as possible; (6) long, persistent action, *i.e.* good chemical stability; and (7) low price.[10] Everything has its limitations; by today's standards, Paul Müller clearly overlooked two other qualities, which are: effective degradation in the environment and no accumulation in the biota. Indeed, with the passing of time, some serious problems relating to the human biologic chain caused by the application of DDT became apparent due to its non-degradable quality. In 1962, Rachel Carson's remarkable book "Silent Spring" was published, which alerted the public to the potential problems of pesticide misuse and promoted modern environmental protection. Today, the basic features of an ideal pesticide should be high activity and selectivity, low mammalian toxicity and non-residue. Therefore, DDT was prohibited many years ago. However, in some developing countries, DDT has been applied because of its high activity, low price and no patent problems. Arguments about pesticides never stop, but the fact is that we cannot give them up. Jerry Cooper[12] suggested that there were 26 primary benefits and 31 secondary benefits of pesticides to mankind and the environment, which are expressed in three aspects: controlling agricultural pests (including diseases and weeds) and vectors of plant disease; controlling human and livestock disease vectors and nuisance organisms; and preventing or controlling organisms that harm other human activities and structures. However, their great contribution to humans is often not properly credited due to the widespread negative evaluations of pesticides. Indeed, the tendency of increasing the production of food through the use of more pesticides is likely to further damage the environment and degrade food and water quality. This does not automatically imply that agrichemicals are useless or totally harmful, but current problems call for much better control of their registration and uses. Eddleston[13] said in his paper that 300 000 people die each year from pesticide self-poisoning in the rural developing world where pesticides are widely used in smallholder agricultural practice. Wilson *et al.*[14] wrote a paper titled "Why farmers continue to use pesticides despite environmental, health and sustainability costs", which expresses the ambivalence. He suggested that the use of chemical inputs such as pesticides has increased agricultural production and productivity. At the same time, negative externalities from such use have increased too. These externalities include damage to agricultural land, fisheries, fauna and flora. Another major externality was the unintentional destruction of beneficial predators of pests by increasing the virulence of many species of agricultural pests. However, despite these high costs, farmers continue to use pesticides in most countries in increasing quantities. Therefore, their control may require the ban of persistent chemicals, education of farmers and rural workers, and close monitoring of residues in the environment and in foodstuffs. To our satisfaction, environmental awareness is growing. Last spring, The European Commission adopted a proposal (Regulation (EU) no. 485/2013) to restrict the use of three pesticides belonging to the neonicotinoids family (clothianidin, imidacloprid and thiamethoxam, Figure 7.2) for a period of 2 years (http://ec.europa.eu/food/animal/liveanimals/bees/neonicotinoids_en.htm), as this group of newly commercialized and widely used insecticides is suspected to harm bees, butterflies, and other non-target species.

7.1.4   Biotechnology in Agriculture

Besides agrichemicals, another way to improve crop production and quality is through the application of novel plant biotechnology, *i.e.* genetically modified organisms (GMOs). The definition of genetic modification is introducing artificial purified genes (clone) or modified genes (extrinsic source) into some target crops, resulting in heritable changes in crops characteristics same as genetic engineering or gene recombination. The first GM crop in the world was the anti-herbicide tobacco invented by Washington University in St. Louis and Monsanto Company in 1983. Because the anti-herbicide gene was introduced into tobacco, the GM tobacco was immune to herbicides. Greater yield was obtained than before. The first field test of GM anti-pesticides and anti-herbicides cotton was carried out in 1986. Commercial GM crops and some products from GM crops came onto the market in the United States in 1996. Since then, biotechnology has been widely applied

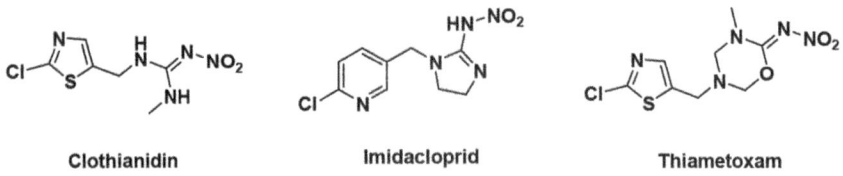

Clothianidin   Imidacloprid   Thiametoxam

Figure 7.2   Structures of clothianidin, imidacloprid and thiametoxam.

in agriculture. The planting area for GM crops has increased from 1.7 million hectares in 1996 to 160 million hectares in 2011, a 94-fold enhancement in 15 years. In 2011, about 100 million people planted GM crops, mainly maize, soybean, and cotton, which laid the foundation for its further development due to its great economic benefits and sustainable environmental effects. ISAAA (International Service for the Acquisition of Agri-biotech Applications) statistics show that planting anti-pesticides GM crops during 1996 to 2008 reduced the use of pesticides by up to 359 million kilograms.

China's study of GM technology in agriculture started in the 1980s and now has formed an integrated GM technology R&D system.[15] Furthermore, China has been leading the world in anti-pesticides cotton and rice studies. A report showed that biotechnology can be used as an alternative to chemical pesticides in a case study of Bt (*Bacillus thuringiensis*) cotton in China.[16]

Although the introduction of new input traits that benefit the growers will continue, it is expected that a second generation of plant biotechnology dealing with output traits that directly benefit the consumers will eventually become prominent. Desirable output traits include prolonging shelf-life through vegetables that are more resistant to rotting and fungal infections, nutritionally superior foods, such as the vitamin A enriched 'golden rice', and agricultural products of medical value, such as edible vaccines. The long lasting acceptance of genetically modified crops might well depend upon the emergence of this second generation of plant biotechnology.[17]

Science is a double-edged sword. Genetic modification has been accompanied by ethical concerns from the beginning. An open letter from the journal Science initiated the discussion in the USA, which culminated in the famous letter by Berg on 26 July 1974.[18,19] People are often in awe of nature. Although GM crops have some advantages over conventional crops, there were worries that the study of new species might strike our traditional moral principles if these technologies are used indiscriminately.[20] GM crop technology is a new frontier. These molecular biology techniques allow the transfer of genes from one organism to another entirely different species without sexual reproduction. This process can be introduced into plant genomes in a controlled manner than can be achieved through conventional breeding and selection of crops.[21] Not surprisingly, the public's opinion of GM crops is not uniform. In the U.S., a lot of consumers continue to gobble their genetically modified corn chips without a second thought. In Europe, however, many consumers have deep objections to GM crops. They think the only benefit is to companies because they sell their expensive seeds; the contemptible advantages are that they could increase yields and perhaps lower pesticide use. Indeed, in an article entitled: "The GM genocide: Thousands of Indian farmers are committing suicide after using genetically modified crops" Prince Charles claimed that thousands of Indian farmers were killing themselves after using GM crops.[22] So, Prince Charles was branded a scaremonger for his articles protesting GM crops from 1998 to 2005.

Although there is no direct evidence to testify that GM products have some harmful effects on humans and the environment, some frequently appearing food safety incidents also made people worry about the potential risks of GM crops. Some people are willing to accept new animals and crops from natural evolution but not from artificial recombinant of DNA from different species.[23] They think that God is being replaced by technology and that punishments from God might fall on them some day. There is no doubt that these debates will continue for a long time.[24]

7.2 Chemical Promoted Biotechnology: Plant Activators (Ecological Pesticides)

7.2.1 Introduction

According to Gavrilescu *et al.*,[25] biotechnology is the application of scientific and engineering principles to the processing of materials by biological agents. Some of the defining technologies of modern biotechnology include: genetic engineering; culture of recombinant microorganisms or cells of animals and plants; metabolic engineering; hybridoma technology; bioelectronics; nanobiotechnology; protein engineering; transgenic animals and plants; tissue and organ engineering; immunological assays; genomics and proteomics; bioseparations; and bioreactor technologies. In a broad sense, biotechnology encompasses all "bioscientific activities".

By these definitions, biopesticides and biofungicides (or microbial pesticides and microbial fungicides) are any biological agents or products derived from microorganisms to be used for controlling insects and weeds in crop protection, and so are classed as biotechnology products. In fact, biopesticides' mechanism for controlling insects is the same as conventional chemical pesticides; their operation targets are pests and fungi. Biopesticides generally tend to be highly target specific, do not leave toxic residues, reduce the risk of resistance development in the target species and have less overall impact on the environment than conventional chemical pesticides. Biofungicides have been used in both the phylloplane and rhizosphere to suppress fungal infection in plants. Species of *Bacillus* and *Pseudomonas* have been successfully used as seed dressings to control certain soil borne plant diseases.[26] Some of the commercial biopesticide products being marketed for use against soil borne plant pathogens have been reviewed.[27] Biopesticides have had some success, but there are concerns relating to their effectiveness and high cost.[28] Today, biopesticides have gained less than 2% of the global pesticides market but this is expected to increase significantly in the future.

Now, pesticides and plant biotechnology are the main crop protection methods to improve production, but worries and disputes over the two methods persist. Is there any other method that can be used for crop protection?

When discussing biotechnology in agriculture, it is apparent that the most common meaning of it is gene-related technology represented by GM crops. In fact, agricultural biotechnology also includes: (1) conventional plant breeding; (2) tissue culture and micropropagation; (3) molecular breeding or marker assisted selection; (4) genetic engineering and GM crops; and (5) molecular diagnostic tools. According to Gavrilescu's definition of biotechnology, the use of a small chemical compound to regulate a general bioprocess also falls within the boundary of chemical biotechnology.

In general, chemical biotechnology is the use of chemically promoted biotechnology methods, such as microbial fermentation and enzyme catalysis with the help of chemicals or additives, and often the use of engineered enzymes to solve chemical problems. The focus is chemical problems, such as syntheses of fuels, chemical feedstocks or polymers from renewable resources, and more environmentally friendly syntheses of enantiomerically pure pharmaceutical intermediates.[29,30] In fact, the methodology of chemical biotechnology involves using chemicals, *i.e.* small organic compounds, which do not have a direct biological activity or function but act as initiators or activators, to make the original bioprocess more selective and efficient in order to solve some biological problem. These small organic compounds interact with biological macromolecules and affect cell physiology and functions, which is currently accepted as a broader area of technology that deals with diverse aspects at the border between chemistry and biology.[31]

Plant activators are known to initiate systemic acquired resistance of plants against a broad spectrum of diseases by adjusting some cascade changes including metabolism or pathogen-related (*PR*) protein expression, which we refer to as chemical biotechnology. Neither plant activators nor their metabolites exhibit antimicrobial activity. An ideal plant activator is expected to display: (1) no direct toxicity to pathogens; (2) no toxicity to plants and animals; (3) no negative effects on plant growth, development and yield; (4) broad spectrum of defense; (5) low loading amount; (6) long lasting protection; (7) low economical cost for farmers; and (8) good profit for producers.[32] Therefore, plant activators are named called crop protection products or "Ecological Pesticides", differentiating them from conventional agrichemicals.

7.2.2 Plant Activators: Action Mechanism

There are two key aspects for a well growing plant. One is external factors including environmental conditions and disease resistance methods; the other is the integrated immune system of the plant. The plant immune system consists of two interconnected tiers of receptors, one outside and one inside the cell. Both systems perceive the invaders, then "inform" the whole plant and sometimes the neighboring plants that the intruder is present by an optional signal. The two systems belong to different classes of plant receptor proteins that detect different types of pathogen molecules.[33]

The first tier is primarily controlled by pattern recognition receptors (PRRs) that are activated by recognition of evolutionarily conserved pathogens or microbial-associated molecular patterns (PAMPs or MAMPs, here P/MAMP). Activation of PRRs leads to intracellular signalling, transcriptional reprogramming and biosynthesis of a complex output response that limits colonization. The system is known as PAMP-triggered immunity (PTI).[34] The second tier is effector-triggered immunity (ETI), which consists of a series of leucine-rich repeat domains (LRRs), the nucleotide-binding LRRs (NLRs). The process runs within the cell, the presence of specific pathogen effectors activates specific NLR proteins that limit pathogen proliferation. On the basis of amino acid sequence and properties, *PR* (pathogenesis-related) proteins are classified into 14 families, *PR-1* to *PR-14*, detected in tobacco, tomato, cucumber, parsley, radish, *Arabidopsis*, and barley in different amounts. Families from *PR-2* to *PR-14* have been characterized as responsible for a specific function or enzymatic activity, including a β-1,3-glucanase (*PR-2*), chitinases (*PR-3*, *PR-4*, *PR-8*, *PR-11*), and a peroxidase (*PR-9*).[35]

Biological changes resulting from receptor responses include ion channel off–on, oxidative bursts, cellular redox changes and protein kinase cascades, which control cellular responses, such as cell wall reinforcement or antimicrobial production, or activate changes in gene expression that then enhance other defensive responses of the plant immune system.[33]

Plant protection resulting from pathogen induced disease-resistance occurs in two ways: the preformed mechanisms and the infection-induced responses of the immune system. In a susceptible plant, disease resistance is able to reduce the pathogen growth on or in the plant, while plants with disease tolerance exhibit little disease damage despite substantial pathogen levels. Disease outcome is determined by the interaction between the pathogen, the plant and the environmental conditions. Except for genetic factors, every living thing in the same environmental conditions displays quite different disease resistance. The main reasons are attributed to the suitable activation of their immune system. In plants, there are two disease resistance systems described in the literature: systemic acquired resistance (SAR) or induced systemic resistance (ISR).[36,37]

Systemic acquired resistance (SAR) is typically activated in healthy systemic tissues next to infected plants. Upon pathogen infection, a mobile signal travels through the vascular system to activate defense responses in distal tissues. Salicylic acid (SA) is an essential signalling molecule for the onset of SAR, which results from the activation of a large set of genes that encode *PR* proteins with antimicrobial activities. Their induction in local as well as in distal parts of the injured or infected plant tissues and their association with the development of SAR suggest a probable role in this type of resistance. Induced systemic resistance (ISR) is activated upon colonization of plant roots by beneficial microorganisms. Like SAR, a long distance signal travels through the vascular system to activate systemic immunity in aboveground plant parts. ISR is commonly considered to be regulated by jasmonic acid (JA)- and ethylene (ET)-dependent signalling pathways. The distinct difference from SAR is that ISR is not associated with the direct activation of *PR* genes. Instead, ISR in plants is controlled by accelerating JA- and ET-dependent gene expression confirmed after pathogen attack.[37,38] Both SAR and ISR are effective against a broad spectrum of virulent plant pathogens.

The resistance in plants induced by pathogens was first recognized in 1901 by Chester,[39] who confirmed those studies by summarizing field observations. Convincing evidence came as late as the 1960s, when a reproducible model of the tobacco plant was developed.[40] The proposed theory has laid the foundation for the preservation of plants in a green manner.

Next, we will discuss the mechanisms of plant activators with SA and JA as examples. In 1979, White[41] showed that the exogenous application of salicylic acid (SA) and some other benzoic acid derivatives induced both resistance to TMV and the accumulation of *PR*-proteins.[42] In more recent studies, SA was found in the phloem of cucumbers infected by either tobacco necrosis virus or *Colletotrichum lagenarium*.[43] TMV treatment of tobacco can also induce the accumulation of endogenously synthesized SA.[44] The levels of endogenous SA have been closely correlated to the induction of the *PR* proteins.[45,46] Moreover, exogenous application of SA induces the same sets of SAR genes that are expressed following biological SAR induction.[47,48] These observations have clearly shown that SA is an endogenous signalling molecule responsible for triggering resistance.

The involvement of SA in SAR has been clearly demonstrated recently with transgenic tobacco that expresses the salicylate hydroxylase gene (*NahG*) from *Pseudomonas putida*.[49] This enzyme catalyzes the conversion of SA to catechol, which has no SAR-inducing activity. The *NahG* expressing plants do not accumulate SA in response to pathogen infection and show no SAR gene expression in the distal portions of the plant. Therefore, SA plays a central role in SAR-signal transduction after pathogen infection, and has long been known as an exogenous inducer of *PR* protein accumulation and resistance.[41]

Now, the question might be asked as to whether SA is able to be used as a candidate for an exogenous applied SAR activator or a practical plant activator. The application of SA can induce local resistance to a broad spectrum of pathogens, as seen with TMV-induced SAR in tobacco. In addition, the same set of genes is induced in the treated tissue to levels comparable to those observed with TMV-induced SAR. Furthermore, neither SA nor its metabolites have significant antibiotic activity. Thus, SA fulfils the criteria of a plant activator. However, there are some concerns with SA. First, although SA is clearly needed to initiate SAR, it is not the translocated signalling molecule. More importantly, reports of SA-mediated resistance are restricted to effects in the treated tissue, indicating that SA does not translocate efficiently when applied exogenously. These SA conjugates lack the phloem mobility of free salicylate. The inability to deliver significant amounts of free SA to systemic tissue after exogenous application seriously hampers the usefulness of SA as an exogenous activator. Finally, severe crop tolerance problems also limit SA use. There is only a narrow safety dose range between efficacious and strongly phytotoxic. So, these limitations do not make SA a practical solution to disease control.

Besides SA, jasmonic acid (JA) and its methyl ester are involved in ISR as signalling molecules. JA and its methyl ester have been mainly considered as mediators of plant responses triggered by wounding and insect feeding, but their involvement in resistance against pathogens has also been proven. For example, methyl jasmonate was used in parsley cell suspension cultures resulting in enhancing phenylpropanoid production as fungal elicitation.[50] This effect was analogous to those observed after pre-treatments with SA or INA (2,6-dichloroisonicotinic acid), but most studies on plants give evidence that the two pathways mediated by SA and JA are clearly distinct, antagonistic or additive.[51]

To summarize, SAR or ISR are broad, physiological immunity that results from infection by pathogens and beneficial microorganisms. Induction of SAR and ISR depend on the plant hormones SA and JA. To make use of it for crop protection, the development of synthetic plant activators is important.

7.3   Practical Plant Activators

7.3.1   Introduction

To be a practical plant activator in crop protection, a compound must fulfil the preliminary condition of no antimicrobial activity *in vitro* either from the compound itself or its possible metabolites. Although the first of these criteria can be easily verified by *in vitro* tests, the second is comparatively difficult to study. The antifungal effect of Fosetyl-Al (CAS no. 39148-24-8) (Figure 7.3) was considered as a resistance reaction until it was proved that its metabolite H_3PO_3 displayed the fungi toxic.[52] In another example, metalaxyl (CAS no. 57837-19-1) (Figure 7.3) was suggested to result in the inhibition of *Phytophthora megasperma* in soybeans, while in fact the defense response resulted from the production of glyceollin behaved the inhibition.[53]

There are two kinds of SAR activator: biotic and abiotic. Biotic activators include extracts from plants and microbes. For example, excellent control of powdery mildews was observed by application of extract of *Rheynoutria sachalinensis*; the extracts from *Bacillus subtilis* have been shown to induce resistance in barley, especially against powdery mildew.[54] It was reported that some plant growth-promoting rhizobacteria may be able to protect plants from foliar diseases when used as a seed treatment or by seed soaking. A strain of *Pseudomonas* was found to be able to protect cucumber against broad spectrum of diseases.[55–57] Chitosan[58,59] and laminarin[60,61] are two typical resistance-related biotic molecules. Biotic plant activators harpin[62] and ComCat®[63] have been commercialized.

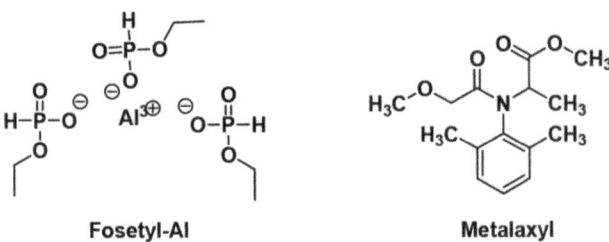

Figure 7.3   Structures of fosetyl-Al and metalaxyl.

Here our focus is on small-molecule abiotic SAR activators. In general, there are two kinds of chemical plant activators: natural and synthetic. Natural plant activators include primary or distant resistance-activating signal molecules, such as salicylic acid (SA),[64] MSA,[65,66] jasmonic acid (JA),[67] MJA[68,69] and abscisic acid (ABA)[70] (Figure 7.4).

Some synthetic or natural exogenous compounds have been proved to potentiate defense responses (Figure 7.5). These include 2,2-dichloro-3,3-dimethylcyclopropane carboxylic acid (DDCC),[71] which is reported to protect

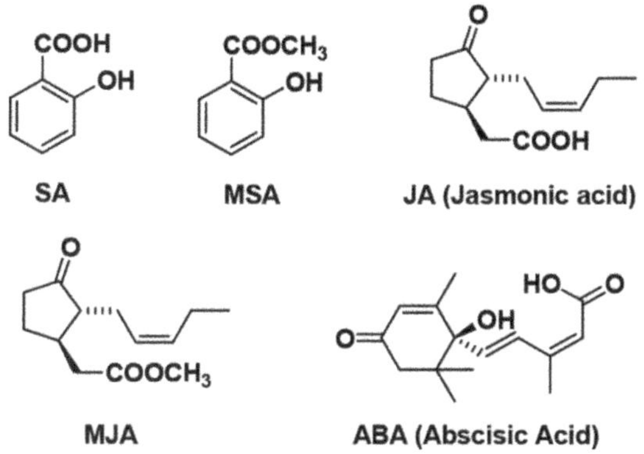

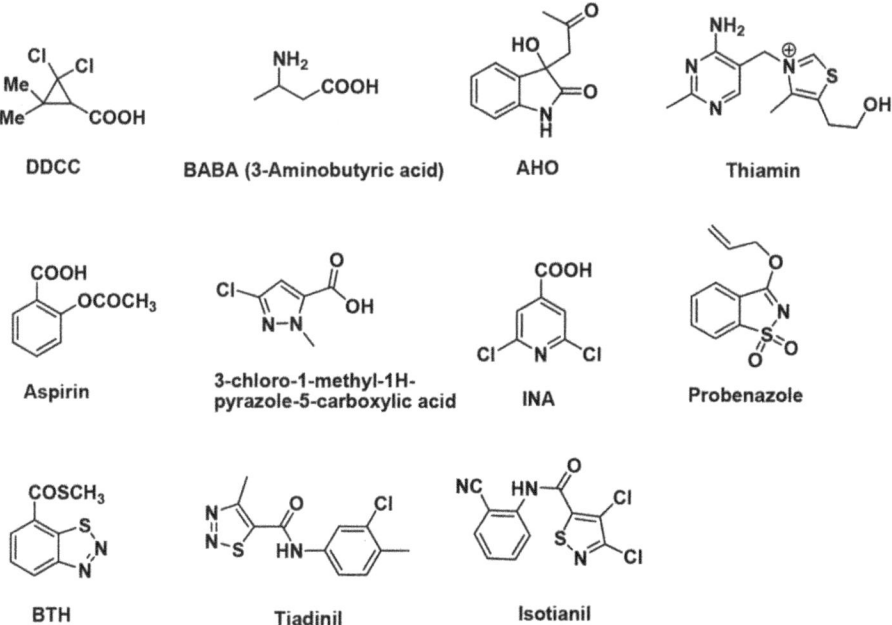

Figure 7.4   Natural activators of resistance in plants.

Figure 7.5   Some synthetic and natural exogenous resistance activators.

rice against *Magnaporthe grisea*. In addition to SAR activators, a synthetic amino acid, β-aminobutyric acid (BABA), was reported to induce disease resistance in *Arabidopsis* by activating a signalling pathway, which is different from SAR and ISR.[72-74] AHO is a natural inducer of SAR represented by an isatin derivative (3-hydroxy-3-(2-oxopropyl)-1*H*-indol-2-one), which was isolated from extracts of the ornamental *Strobilanthes cusia* and found to induce resistance in plants to a broad range of diseases.[75] Abdel-Monaim[76] demonstrated that thiamine gave the best results under greenhouse and field conditions. The activity of defense-related enzymes, including peroxidase (PO), polyphenol oxidase (PPO), phenylalanine ammonia lyase (PAL), and pathogenesis related (*PR*) protein (chitinase), were increased in the inoculated and non-inoculated plants treated with thiamine. The efficiency of acetylsalicylic acid (ASA or aspirin) in inducing localized acquired resistance against infection by *Erwinia carotovora* subsp. *carotovora* was evaluated by treating potato tubers with ASA. Statistical analysis revealed that treatment of tubers by immersion in ASA solutions at low concentrations induced a significant reduction in the incidence of soft rot.[42] Nishioka[77] described the synthesis and evaluation of a novel class of plant activator 1*H*-pyrazole-5-carboxylic acid derivatives. Structure and activity relationships illustrated that the 5-carboxyl group played an important role in the activity and that a halogen atom such as a chlorine or bromine at the 3-position further enhanced the activity. Among the derivatives, 3-chloro-1-methyl-1*H*-pyrazole-5-carboxylic acid (CMPA) was the most potent against rice blast with an ED_{80} (80% effective dose, the concentration needed for 80% inhibition of disease development) of 0.05 mg per pot. CMPA did not have any significant effects on the hyphal growth, spore germination, and appressorium formation of *Pyricularia oryzae*, suggesting that CMPA indeed induced systemic acquired resistance in rice plants. Yasuda *et al.*[78] suggested that CMPA was an effective SAR inducer in tobacco; soil drench application of CMPA induced *PR* gene expression and a broad range of disease resistance. Both analysis of CMPA's effects on *NahG* transgenic tobacco plants and SA measurement in wild-type plants indicated that CMPA-induced resistance enhancement did not require SA. Therefore, CMPA induced SAR by triggering the signalling in the same mechanism as synthetic SAR reducers benzo-1,2,3-thiadiazole-7-carbothioic acid *S*-methyl ester (BTH) and *N*-cyanomethyl-2-chloroisonicotinamide (NCI).

Next, we will introduce several practical or commercial plant activators. These synthetic plant activators include BTH,[79] probenazole,[80] INA,[81] TDL,[82] and some BTH analog trifluoroethyl esters that have been developed at our group.[83,84]

7.3.2   BTH and Its Analogs

BTH (CAS no. 135158-54-2, chemical name *S*-methyl benzo-1,2,3-thiadiazole-7-carbothioate, or acibenzolar-*S*-methyl [ASM]), was introduced by Ciba–Geigy AG (now Syngenta AG) in Germany and Switzerland in 1996 (Patent EP 313512, Manufacturers Syngenta). It was registered in the USA as

Actigard and in Europe as Bion.[85] It was the first synthetic commercial plant activator from an unexpected product in the preparation of sulfonylurea herbicides. In the screening for bioactivity, compound CGA 245704 was discovered to trigger SAR in plants.

Structure–activity relationship analysis of a large number of analogs showed that the activity was substantially restricted to the benzothiadiazole skeleton bearing a carboxyl group, either free or esterified, on the 7-position (see Figure 7.6). The phenyl ring could be replaced by a pyridine one as long as the N was located in suitable positions: It may replace 4-CH (Compound A) or 6-CH (Compound B) but not the 5-CH position.[86] Because of difficulties in the preparation of pyridine-analogs, BTH was selected to be developed.

Encouraged by the success of BTH, a large number of benzothiadiazole derivatives were developed.[87-90] New synthetic elicitors that showed powerful eliciting activities on taxoid biosynthesis by *Taxus chinensis* cell suspension cultures were obtained as shown in Figure 7.7.[83] JA and JA derivatives

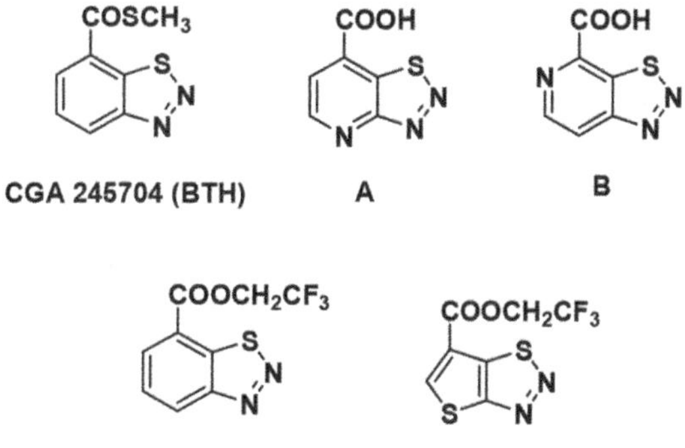

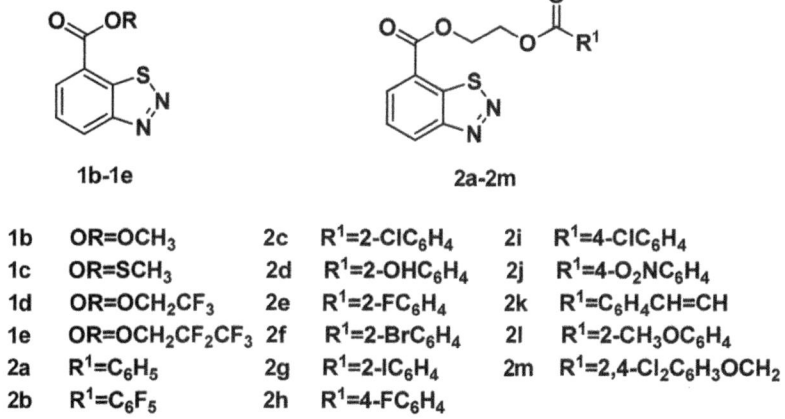

Figure 7.6 BTH and its derivatives as plant activators.

1b	OR=OCH$_3$	2c	R^1=2-ClC$_6$H$_4$	2i	R^1=4-ClC$_6$H$_4$
1c	OR=SCH$_3$	2d	R^1=2-OHC$_6$H$_4$	2j	R^1=4-O$_2$NC$_6$H$_4$
1d	OR=OCH$_2$CF$_3$	2e	R^1=2-FC$_6$H$_4$	2k	R^1=C$_6$H$_4$CH=CH
1e	OR=OCH$_2$CF$_2$CF$_3$	2f	R^1=2-BrC$_6$H$_4$	2l	R^1=2-CH$_3$OC$_6$H$_4$
2a	R^1=C$_6$H$_5$	2g	R^1=2-IC$_6$H$_4$	2m	R^1=2,4-Cl$_2$C$_6$H$_3$OCH$_2$
2b	R^1=C$_6$F$_5$	2h	R^1=4-FC$_6$H$_4$		

Figure 7.7 Synthetic BTH derivatives.

are effective elicitors in plant cell culture to enhance secondary metabolites, as described in Chapter 4. It is the resistance response induced by elicitors that initiates the changes in metabolism. Therefore, a series of BTH derivatives with eliciting activity were developed (Figure 7.7).[83] For example, benzo-1,2,3-thiadiazole-7-carboxylic acid 2-(2-hydroxybenzoxyl) ethyl ester (**2d**) was more effective and resulted in a nearly 40% increase in taxuyunnanine C content and production in comparison with methyl jasmonate, which was previously reported as the most powerful chemical elicitor for taxoid biosynthesis. The novel class of benzo-1,2,3-thiadiazole-7-carboxylic acid elicitors was found to induce plant defense responses, including promotion of H_2O_2 levels originating from oxidative burst and activation of phenylalanine ammonia lyase. Interestingly, the plant defense responses corresponded well to the superior stimulating activity in *T. chinensis* cell cultures. The work indicated that the newly synthesized benzothiadiazoles could act as a new family of elicitors for taxoid biosynthesis in plant cells[83] (Figure 7.7).

In order to develop novel plant activators, the series of benzo-1,2,3-thiadiazole-7-carboxylate derivatives (Figure 7.7) was evaluated.[91] Most of them showed good activity, especially fluoro-containing **1d** and **1e**, which displayed excellent SAR-inducing activity against cucumber *Erysiphe cichoracearum* and *Colletotrichum lagenarium* in assay screening. Field test results illustrated that compounds **1d** and **1e** were more potent than the commercialized acibenzolar-*S*-methyl (BTH) towards the test diseases. Furthermore, preparation of compound **1d** was more facile and cheaper than BTH, which would be helpful for further application in agricultural plant protection.

Field tests of compounds **1d** and **1e** were carried out and the results for efficiency and disease index obtained from eqn (7.1) and (7.2) are displayed in Tables 7.1 and 7.2. From every subarea, 10 plants were chosen and their

Table 7.1 Field test results of **3d** and **3e** against *Pseudoperonospora cubensis* on cucumber.[a]

Compound	Concentration (mg L^{-1})	Disease index	Efficiency (%)
3d	500	10.07	76.44
	250	13.81	67.54
	100	14.05	67.12
3e	500	10.97	74.34
	250	12.40	70.98
	100	12.49	70.78
3c	250	14.02	67.19
Solvent	ND[b]	37.05	13.31
50% dimethomorph (WP)	500	5.94	86.09
CK		42.74	ND

[a]Field tests in three locations were consistent and only representative data is shown, which is from tests carried out in Peking, 2008. **3c** is the BTH prepared in the laboratory.
[b]ND = not detected.

leaves were investigated and classified to different levels according to the following standards:

0: No lesions;
1: Lesions occupied less than 5% of the leaf area;
3: Lesions occupied 6–10% of the leaf area;
5: Lesions occupied 11–25% of the leaf area;
7: Lesions occupied 26–50% of the leaf area;
9: Lesions occupied more than 50% of the leaf area.

The disease index and the inhibition efficiency were calculated according to eqn (7.1) and (7.2), respectively.

$$I = \frac{\sum(Y \times C)}{N \times 9} \times 100 \tag{7.1}$$

$$\text{Efficiency} = \frac{I_c - I_a}{I_c} \tag{7.2}$$

where I is the disease index, Y is the number of leaves of a certain class, C is the corresponding lesion level, and N is the total number of leaves of all levels; I_c is the disease index of CK plants and I_a is the disease index of plants with test compounds applied.

Recently, some thieno[2,3-d][1,2,3]thiadiazole-6-carboxylate derivatives (**3c**) have also been synthesized as bio-isosteric models of BTH,[92,93] but no bioassay data were provided in the paper. In our laboratory, some thieno[2,3-d][1,2,3]thiadiazole-6-carboxylate derivatives were obtained by ester modification (Figure 7.8).[94] A few plant activators that are more potent than BTH were obtained in primary screening by introduction of fluoro-containing groups. The *in vivo* bioassay showed that these novel compounds had good efficacy

Table 7.2 Field test results of **3d** and **3e** against *Erysiphe cichoracearum* on cucumber.[a]

Compound	Concentration (mg L^{-1})	Disease index	Efficiency (%)
3d	500	1.39	94.50
	250	1.88	92.55
	100	1.92	92.39
3e	500	1.10	95.65
	250	1.38	94.53
	100	3.28	87.02
3c	250	2.38	90.58
Solvent	N.D.[b]	22.27	11.87
12.5% myclobutanil (EC)	500	2.89	88.56
CK		25.27	N.D.

[a]Field tests in three locations were consistent and only representative data is shown, which is from tests carried out in Peking, 2008. Efficiency is the average of all replicates.
[b]N.D. = not detected.

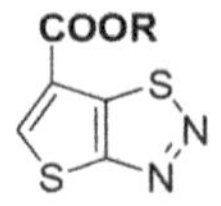

Compds.　R

COOR

3a　CH_3

3c　CH_2CF_3

3f　$CH_2CF_2CF_3$

3b-3m　3g　$CH(CF_3)_2$

Figure 7.8 Thieno[2,3-*d*][1,2,3]thiadiazole-6-carboxylate derivatives.

Table 7.3 *In vivo* induction activity of the target compounds.[a]

Compound	Efficacy (%)						
	MM	CC	PL	PI	RS	FO	BC
1a	90 ± 3	77 ± 3	42 ± 4	81 ± 5	-6 ± 1	5 ± 4	37 ± 4
1c	69 ± 6	52 ± 5	42 ± 4	67 ± 6	-6 ± 3	7 ± 3	41 ± 5
1f	51 ± 4	58 ± 3	48 ± 5	74 ± 4	0	9 ± 1	32 ± 3
1g	62 ± 2	59 ± 7	41 ± 2	40 ± 3	-1 ± 2	14 ± 4	10 ± 1
BTH	88 ± 3	-8 ± 4	76 ± 8	65 ± 5	16 ± 4	19 ± 2	47 ± 6
50% kresoxim-methyl (WG)	98 ± 1						
75% chlorothalonil (WP)		90 ± 3					
20% bismerthlazol (WP)			68 ± 5				
50% dimethomorph (WP)				97 ± 2			
5% validamycin A (WP)					83 ± 6		
70% mildothane (WP)						75 ± 5	
50% procymidone (WP)							42 ± 5

[a]BTH was synthesized in our lab; micro-organisms used: *Mycosphaerella melonis* (MM), *Coryne-spora cassiicola* (CC), *Fusarium oxysporum* (FO), *Botrytis cinerea* (BC), *Pseudomonas syringae* pv. *lachrymans* (PL), *Phytophthora infestans* (PI) and *Rhizoctonia solani* (RS); efficiency is the average of all repeated tests; MM, CC, BC, FO and PL were tested on cucumber seedlings; PI was tested on tomato seedlings; RS was tested on rice seedlings.

against seven plant diseases (Table 7.3). Especially, compounds **3a** and **3c** were more potent than BTH. Almost no fungicidal activity was observed for the active compounds in *in vitro* assay, which matched the requirements for plant activators.

The main application of BTH is to control powdery mildew in wheat and barley.[95] BTH is an important defense response inducer,[96] but it did not affect the pathogen metabolism like fungicides. In order to determine whether BTH played a role in SAR, a susceptible wheat cultivar was infected with *Erysiphe graminis* f. sp. *tritici.*, then sprayed with BTH ten days after inoculation. A hypersensitive response was successfully initiated. This was a clear indication that BTH indeed activated SAR to induce a defense response.[97] While the defense response was initiated without the production of SA,[78] the BTH-initiated response did, however, trigger a similar downstream signal pathway to SA.

Arabidopsis plants treated with BTH induced the expression of *PR-1*.[98] van Hulten *et al.*[99] also indicated that treated *Arabidopsis* plants with BTH induced *PR-1* defense gene expression after *Hyaloperonospora parasitica* inoculation, which led to reduced pathogen colonization. BTH did, however, not regulate all *PR*-defense genes in the same way. Once *Brassica oleracea* seedlings were sprayed with BTH, induction of β-1,3-glucanase activity and *PR-2* gene expression was observed, while *PR-1*, *PR-3*, *PR-5* expression and chitinase activity were unaffected.[100] Another example where BTH induced a defense response was in wheat. A study showed increased resistance against powdery mildew (*Blumeria graminis*), leaf rust (*Puccinia triticina*) and *Septoria* leaf spot (*Septoria* spp.).[101] BTH also induced a resistance response in peach fruit[102] and Yali pear infected with *Penicillium expansum*.[103] Even though it does not protect all plants against pathogens, the induced defense response was much stronger than responses triggered by SA or JA. The augmented level of resistance of tissues by BTH or MJA-treated seeds was associated with rapid increases in the activity of the *PR* proteins chitinase and peroxidase.[104] MJA also induced a rapid and transient accumulation of lipoxygenase. Moreover, BTH and MJA treatments led to differential induction of particular *de novo* synthesized isoenzymes of these proteins. Results indicate that BTH and MJA applied to melon seeds may activate diverse metabolic pathways in seedlings leading to enhanced resistance against distinct pathogens.

Benzothiadiazole and benzotriazole are both unstable structures and have a tendency to lose N_2. Therefore, BTH might form an SA derivative as shown in Figure 7.9. However, there are few reports on the metabolic pathway. The Australian Pesticides and Veterinary Medicines Authority proposed that the metabolic pathway of acibenzolar-*S*-methyl in wheat, tomato and tobacco proceeds *via* hydrolysis of the parent molecule to the acid metabolite (1) followed by glycosylation with sugars. Subsequent oxidation of the phenyl ring led to the 5-hydroxylated acid metabolite (2) and the 4-hydroxylated acid metabolite (3) (tomato and tobacco) followed by glycosylation to their *O*-glycoside. The metabolic pathway in plants is qualitatively similar in all crops tested (Figure 7.9). Different metabolic pathway in animals (hens and goats) and toxic evaluations are also presented in the literature.[105]

7.3.3   2,6-Dichloroisonicotinic Acid (INA) and Its Derivative (NCI)

INA (Figure 7.10), like BTH, is one of the most important plant activators in initiating of SAR[97] and is able to induce a defense response without SA.[78] Therefore, the action mechanism of INA is confirmed to induce SAR through the same signal transduction pathway as SA.[106] Another similarity between SA and INA is that both consist of a hexagon-shaped ring with a carboxyl

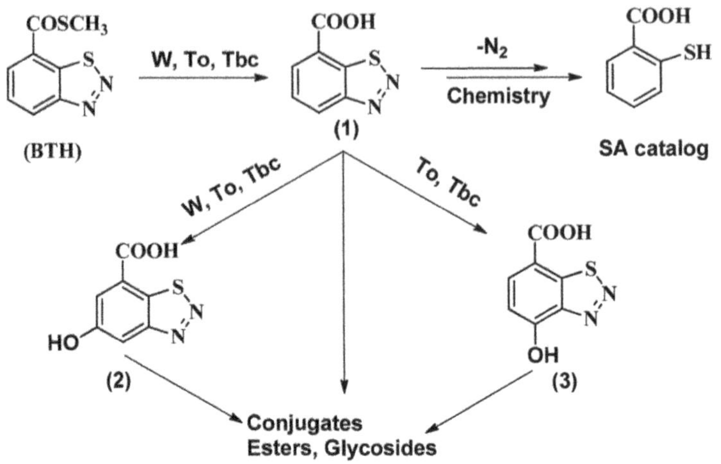

Figure 7.9 Proposed metabolism of BTH in wheat (W), tomato (To) and tobacco (Tbc), and chemically.

Figure 7.10 Structures of INA and analogue NCI with similar activities as inducers of SAR.

group,[107] and it was found that SA was more effective and faster in the activation of a defense response compared to INA. Research indicated, like BTH, that when a *NahG* expressing plant was sprayed with INA, the defense against *P. infestans* reduced.[108] A possible explanation lies in the fact that INA can be considered an analogue of SA. Since plants expressing *NahG* are defective in the SA dependent signalling pathway,[109] *NahG* plants treated with INA will also be unable to induce a defense response. It was also found that INA derivatives act as elicitors that can increase secondary metabolism with the same resistance responses as MJA.[110] In a study done by Umemura *et al.*,[111] rice treated with INA showed a phytotoxic response in leaves, but it is still able to protect a plant against a potential pathogen infection. INA was more efficient in the induced expression of the uridine diphosphate glucose (UDP glucose) and SA glucosyltranferase (*OsSGT1*) gene.

Because of the phytotoxicity of INA, *N*-cyanomethyl-2-chloroisonicotin-amide (NCI) was developed as a synthetic plant activator against rice blast disease[90,112] (Figure 7.10). In evaluation of the structure and activity relationship, the *N*-cyanomethyl substituent was found from screening to more

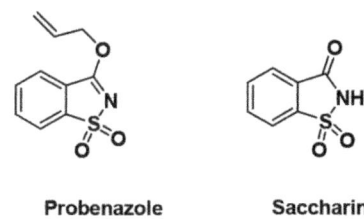

Probenazole   **Saccharin**

Figure 7.11   Structures of probenazole (PBZ) and its active metabolite saccharin (BIT).

effective than other *N*-alkyl groups. Nakashita *et al.*[90] demonstrated that NCI fulfilled all the criteria of an SAR activator in tobacco. Measurement of SA accumulation and use of *NahG* transgenic tobacco plants revealed that NCI-mediated resistance enhancement in tobacco didn't require SA, which suggested that NCI induced disease resistance by triggering the signalling pathway at the same level as BTH and INA.

7.3.4   Probenazole (PBZ) and Saccharin (BIT)

Probenazole (3-allyloxy-1,2-benzisothiazole-1,1-oxide, PBZ) (Figure 7.11) is a plant activator developed by Meiji Seika Kaisha Ltd. for rice blast control. The registered product, Oryzemate® granules, containing probenazole, has been widely used in Japan against rice blast since 1975 because it provides excellent blast control and controls leaf blast for 40–70 days after application.[113] After being applied in rice plants, the whole plant is conferred with systemic resistance transferred from the roots, where PBZ is absorbed.[114] The big advantage of PBZ is that no development of resistance in the target fungus has been observed over many years of use.[115]

There is a delicate relationship between plants and pathogens. In general, when environmental conditions such as temperature and humidity are favourable for the pathogen, the pathogen can easily invade the plant. When the plant defense system works effectively, on the other hand, the plant can defend the pathogen attack. The pivotal role of plant activators is to initiate the plant's defense system. In many cases, PBZ is used as a fungicide, but many studies have demonstrated that PBZ has only weak antimicrobial activity against *Magnaporthe grisea*, although it displays excellent blast control.[116] From experiments on the effect of PBZ on the blast fungus,[117] Iwata concluded that PBZ did not have any fungicidal activity against the fungus; during the process, PBZ did not change into a fungicidal substance within the rice plant tissue or in an environment such as paddy water. The metabolite of PBZ, 1,2-benzisothiazoline-3-one-1,1-dioxide (BIT), a well-known sweetener used in foods and drinks, has almost no biotic toxicity meaning that it and its metabolites do not affect the growth or infectivity of the blast fungus,[118] so BIT is also considered a good SAR elicitor for inducing resistance against a broad spectrum of pathogens in both cereals and leguminous plants.[119]

The phenylpropanoid pathway plays an important role in the plant defense system. When the plant is being infected, lignin is synthesized as a physical barrier against pathogen invasion, while a phytoalexin with antimicrobial activity is produced. These contribute to limiting pathogen invasion in the plant tissue. Iwata[117] showed that PBZ activates the phenylpropanoid pathway and thereby enhances the defense response in the plant. Some fungicidal substances were found to accumulate within the tissue of the treated or inoculated rice leaf. Since PBZ and its metabolites do not have fungicidal activity, these active substances were thought to have originated from the rice plant, which were identified as hydroxyl unsaturated fatty acids derived from α-linolenic acid. These hydroxyl unsaturated fatty acids were proposed to arise from a biosynthesis pathway as follows: first, α-linolenic acid produced from phospholipid in cell plasma membrane by phospholipase A_2 is oxidized into hydroperoxylinolenic acids by lipoxygenase; then rapidly reduced to hydroxides from hydroperoxides. Both enzymes in the rice leaf were enhanced when the plant was inoculated with a resistance activator, incompatible race of the blast fungus, suggesting participation of both enzymes in defense response. The hydroperoxide synthesis forms part of the octadecanoid (18-carbon) pathway in which the plant hormone jasmonic acid, an endogenous activator of defense gene expression and phytoalexin biosynthesis, is synthesized.[120,121]

Extensive studies have demonstrated that PBZ activates some defense responding mechanism of rice.[122] Recently, Yoshioka *et al.*[123] demonstrated that PBZ and its active metabolite BIT induce SAR in *Arabidopsis* by stimulating the site upstream of SA in the SAR pathway.[124] The role of SA in activating resistance following PBZ or BIT treatment was studied using *NahG* transgenic plants. PBZ or BIT treatment did not induce disease resistance and *PR-1* expression in *NahG* transgenic plants, which indicated that these compounds induce SAR by triggering signalling at a point upstream of SA accumulation.[123]

A new rice gene, *PBZ1*, induced by PBZ application was found by Midoh and Iwata when screening for rice expression.[113] The amino acid sequence estimated for the *PBZ1* gene showed about 30% homology with *PR-10* protein. This *PR* protein is induced after a pathogen infection and is thought to be an infection response and defense-participating protein.[115,125,126] Sakamoto *et al.*[127] isolated another rice gene, *RPR1* (rice PBZ responsible gene), by a differential display technique. Recently, researchers have reported that many defense related genes in the rice plant are induced by application of PBZ.[128] The conclusion on the mode of action of PBZ is that it prevents the invasion of pathogens by inducing many defense genes through the signal transduction pathway.[124]

7.3.5   Tiadinil (TDL): Thiazole Derivatives

Tiadinil (3-chloro-4,4-dimethyl-1,2,3-thiadiazole-5-carboxanilide, CAS no. 223580-51-6, Figure 7.12), a novel class plant activator, was discovered and developed by Nihon Nohyaku Co., Ltd. It was registered in Japan in 2003

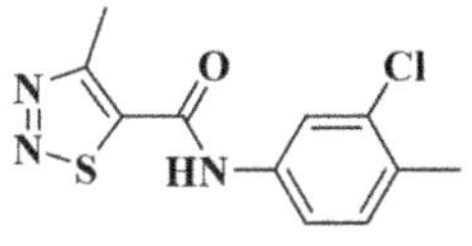

Tiadinil

Figure 7.12   Structure of TDL.

(trade name: V-GET®) and also approved in combination with some insecticides, fungicides, and herbicides.

While evaluating the relationship between activity and structure, Tsubata *et al.*[128] thought that the key chemical structure was a heterocyclic moiety, 1,2,3-thiadiazole, the same moiety as in BTH. A carboxyl group at the 5-position of heterocyclic moiety was considered important. A 4-position substituent alkyl with a chain length of one to three was optimal, and suitable hydrophobicity was also an important factor in the activity. The 5-position of 1,2,3-thiadiazole moiety carboxylic acid derivatives such as esters and anilides showed high activity. It was suggested that the 1,2,3-thiadiazole-5-carboxylic acid produced by hydrolysis was the important moiety. The benzene moiety 3'-chloro-4'-methylanilide showed very good activity under different conditions throughout the selection process used in the development.[129]

TDL exhibited excellent activity against rice blast, and the control activity even lasted more than 80 days after treatment. When TDL was applied at 7–20 days before the first appearance of leaf blast, TDL showed stable control activity and was not affected under various field conditions, such as infection pressure of disease, type of soil, depth of paddy water or depth of transplanted seedlings. TDL exhibited additional control activity against some bacterial diseases in field trials, such as rice bacterial leaf blight and grain rot, which suggested that TDL was adaptable to the simultaneous control of a wide range of rice diseases in practical use.

TDL and its formulation exhibited quite low acute toxicity toward laboratory animals *via* any exposure route. No critical eye or skin irritation, or skin sensitization, was recognized. The toxicity of TDL toward non-target aqueous species (fish, crustacean and algae), earthworms, silk worms and honey bees was also quite low.[128] Metabolism studies employing radiolabelled compounds showed that the compound was rapidly metabolized *via* hydrolysis of the amide bond and oxidation of both the methyl group of the 4-position of the thiadiazole ring and the phenyl ring in plants, soil and animals. No significant translocation of TDL and its metabolites into rice grain was found.

Results from induction of resistance in tobacco against TMV by TDL showed that TDL could induce SAR-like disease resistance by evoking *PR* gene expression in tobacco, the same as SA, BTH and PBZ.[128–130] TDL did not inhibit the morphogenesis of the infection process of rice blast fungus, *Magnaporthe grisea*, although the intracellular hyphal growth of *M. grisea* was remarkably inhibited in the first invaded cell of the inner epidermal tissues of rice leaf sheath excised from TDL-treated rice plants.[131] Moreover, the enhancement of several resistance-related genes reported previously, such as *PBZ1*, *RPR-1*, was detected in TDL-treated rice plants.[82,131] These results strongly suggest that tiadinil provides excellent blast control by activation of the resistance mechanism possessed primarily in the host plant, and works as a plant activator in rice.

Plant activators BTH and TDL both contain the same structural moiety: 1,2,3-thiadiazole. As a class of compounds with various interesting properties, 1,2,3-thiadiazoles are becoming a rapidly growing and independent branch of chemistry. Many 1,2,3-thiadiazoles have shown biological activities, such as antiviral,[132,133] systemic acquired resistance,[91] and fungicidal.[134-136]

Therefore, Fan *et al.*[137-142] synthesized a large number of 1,2,3-thiadiazole derivatives by combining the active substructures of 1,2,3-thiadiazole and thiazole; a series of compounds with high bioassay were obtained. Among them, methiadinil, *N*-(5-methyl-1,3-thiazole-2-yl)-4-methyl-1,2,3-thiadiazole-5-carboxylicamide (Figure 7.13),[143] indeed had excellent systemic acquired resistance against disease caused by fungi, bacteria and virus. Field tests in an agroecosystem validated its efficacy. Preliminary studies on the mode of action by tobacco against TMV, rice against *Magnaporthe grisea* and induction of *Hypericum sampsonii* producing hypericin compounds also validated that its main mechanisms were the same as those of BTH and TDL.

Toxicological studies indicated that methiadinil had low toxicity. It is now under development by a developing pesticide manufacture as a candidate elicitor with independent properties in China.

7.3.6 Plant Defense against Herbivores by Plant Activators

Plant defenses contain constitutive responses and inducible responses. Two functions of the induced defense response of plants are against pathogens directly and against herbivores indirectly. Once plants are damaged by herbivorous arthropods, an induced volatile blend will be emitted so that carnivores can sense it from a distance.[144,145] Indirect defenses work by attracting

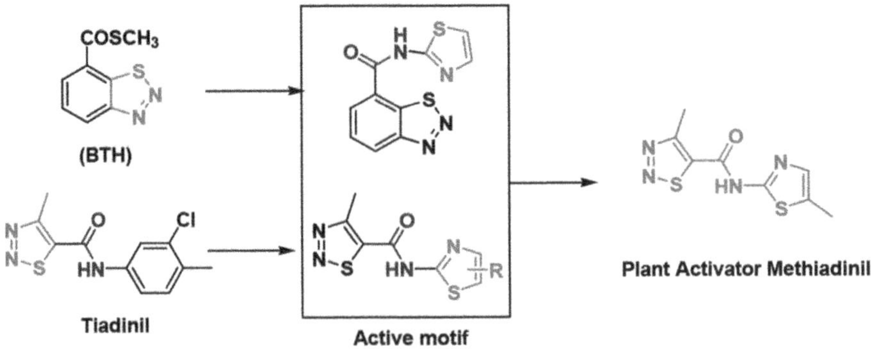

Figure 7.13   1,2,3-Thiadiazole derivatives prepared by Fan's group.

the natural enemies of herbivores, such as parasitoids or predators, actively reducing the number of herbivores. For example, some plants emit specific volatile chemicals that attract natural enemies when they are attacked by herbivores. The SA pathway is mainly activated by pathogens, whereas the JA pathway is activated by herbivores.[146-148] The plant hormone jasmonic acid (JA) plays a central role in plant defense against herbivores. Herbivore damage elicits a rapid and transient JA burst in the wounded leaves and JA functions as a signal to mediate the accumulation of various secondary metabolites that confer resistance to herbivores.[149]

Some exogenous plant activators inducing resistance to herbivores are further examples of chemical biotechnology. MJA,[146,149] BTH,[150,151] TDL[152] and NCI[153] are all reported to be able to enhance herbivore resistance. It was found that treatment with BTH increased the attractiveness of maize plants to parasitic wasp.[60] The parasitic wasps are parasitoids of various animals, mainly other arthropods. Many of them are considered beneficial to humans because they control populations of agricultural pests.

Interestingly, plants treated with the plant activators did not show any consistent increase in volatile emissions. On the contrary, treated plants released less herbivore-induced volatiles such as indole, which has been reported to interfere with parasitoid attraction. The results support the yet undetectable and unidentified phenomena that plant activators are major factors for parasitoid attraction, and these attractants may be masked by some of the major compounds in the volatile blends. This study confirms that activators of pathogen resistance are compatible with the biological control of insect pests and may even help to improve it.

Rostas and Turlings[154] showed that a SAR against pathogens induced by BTH application affected the plant's attractiveness to parasitoids. They found that BTH induced significant systemic resistance in maize seedlings against the pathogen *Setosphaeria turcica*. In addition, BTH treatment prior to *Spodoptera littoralis* caterpillar feeding made the plants far more attractive to the parasitoid *Microplitis rufiventris* than plants that were only damaged by the herbivore, although the volatile emission was not affected by BTH treatment. These findings suggest that if the defense mechanisms of plants can be activated against pathogens and herbivores simultaneously, it would be a useful technique for integrated pest and pathogen management.

7.4 Future Development of Plant Activators

There is no doubt that rapid development of plant activators will come, but the characteristics of no specific target, no practical evaluation methods until now and largely unknown molecular mechanisms of immune induction make it difficult to develop more practical plant activators. High-throughput screening has been widely used in new drug discovery and has become a general research strategy that fits well with the structural diversity of plant activators. Noutoshi *et al.*[155] established a high-throughput screening procedure to screen for compounds that specifically potentiate pathogen-activated

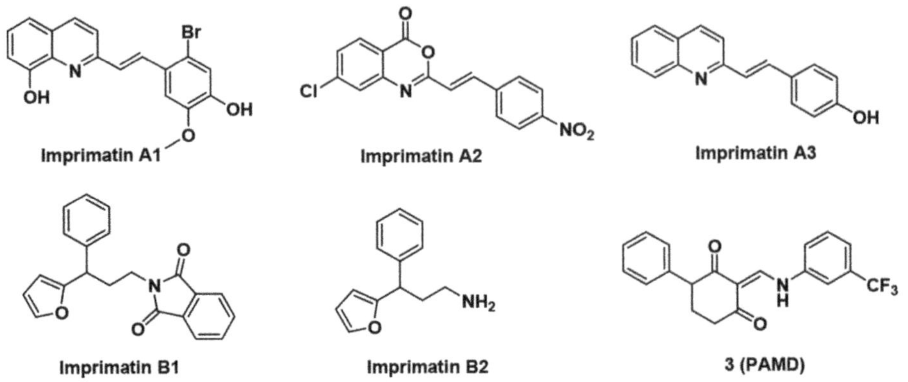

Figure 7.14 Novel SAR biochemicals discovered from high-throughput screening.

cell death in *Arabidopsis thaliana* suspension cultured cells. 5 compounds (Figure 7.14) different from those known SAR activators with priming the immune response were obtained by inducing compounds' inhibition two SA glucosyltransferases (SAGTs) *in vitro* to manipulate the active free SA pool *via* SA-inactivating enzymes.

Seo *et al.*[156] suggested another compound from high-throughput screening of 9600 compounds. 8-day-old transgenic *PR1::GUS Arabidopsis* sprayed with 2 mM SA was treated by novel candidates; the inhibitory effect of compound 3 (PAMD) displayed better potency and will be further studied (Figure 7.14).

In our laboratory, similar high-throughput screening methods by application of ligand-based virtual screening program SHAFTS to discover novel SAR active compounds are ongoing.[157]

In summary, with the development of science and technology, more and more plant activators will be developed as ecological pesticides. Because of environmental factors and ethical considerations, ecological pesticides derived from chemical biotechnology in agriculture will be applied more in the future.

References

1. L. R. Brown, *Who will feed China*, Washington, DC, World Watch Institute, 1995.
2. F. P. Carvalho, *Environ. Sci. Policy*, 2006, **9**, 685–692.
3. N. Alexandratos, *Proc. Natl. Acad. Sci. U. S. A.*, 1999, **96**, 5908–5914.
4. H. Spiertz, *Curr. Opin. Environ. Sustainability*, 2010, 2, 439–443.
5. R. Lal, *Soil Tillage Res.*, 2009, **102**, 1–4.
6. W. M. Stewart and T. L. Roberts, *Procedia Eng.*, 2012, **46**, 76–82.
7. J. J. Schroder, D. Scholefield, F. Cabral and G. Hofman, *Environ. Sci. Policy*, 2004, 7, 15–23.
8. P. M. Rutherford, M. J. Dudas and J. M. Arocena, *Environ. Technol.*, 1995, **16**, 343–354.

9. W. Zhang, F. Jiang and J. Ou, *Proc. Int. Acad. Ecol. Environ. Sci.*, 2011, **1**, 125–144.
10. W. M. Jarman and K. Ballschmiter, *Endeavour*, 2012, **36**, 131–142.
11. J. Zadoks, *Eur. J. Plant Pathol.*, 1995, **101**, 11–13.
12. J. Cooper and H. Dobson, *Crop Prot.*, 2007, **26**, 1337–1348.
13. M. Eddleston and D. N. Bateman, *Medicine*, 2012, **40**, 147–150.
14. C. Wilson and C. Tisdell, *Ecol. Econ.*, 2001, **39**, 449–462.
15. Z.-H. Xu and S.-N. Bai, *Trends Plant Sci.*, 2002, **7**, 374–375.
16. J. Huang, R. Hu, C. Pray, F. Qiao and S. Rozelle, *Agr. Econ.*, 2003, **29**, 55–67.
17. C. McCullum, C. Benbrook, L. Knowles, S. Roberts and T. Schryver, *J. Nutr. Educ. Behav.*, 2003, **35**, 319–332.
18. P. Berg, D. Baltimore, H. W. Boyer, S. N. Cohen, R. W. Davis, D. S. Hogness, D. Nathans, R. Roblin, J. D. Watson, S. Weissman and N. D. Zinder, Potential Biohazards of Recombinant DNA Molecules, *Science, New Series*, 1974, **185**(4148), 303.
19. P. Berg, D. Baltimore, H. W. Boyer, S. N. Cohen, R. W. Davis, D. S. Hogness, D. Nathans, R. Roblin, J. D. Watson, S. Weissman and N. D. Zinder, *Science*, 1974, **185**, 303.
20. H. I. Miller, *New Biotechnol.*, 2010, **27**, 628–634.
21. A. Konig, A. Cockburn, R. W. Crevel, E. Debruyne, R. Grafstroem, U. Hammerling, I. Kimber, I. Knudsen, H. A. Kuiper, A. A. Peijnenburg, A. H. Penninks, M. Poulsen, M. Schauzu and J. M. Wal, *Food and Chemical Toxicology: An International Journal Published for the British Industrial Biological Research Association*, 2004, vol. 42, pp. 1047–1088.
22. P. C. Abhilash and N. Singh, *J. Hazard. Mater.*, 2009, **165**, 1–12.
23. J. L. Domingo and J. Gine Bordonaba, *Environ. Int.*, 2011, **37**, 734–742.
24. S. O. Hansson and K. Joelsson, *J. Agr. Environ. Ethics*, 2012, **26**, 759–770.
25. M. Gavrilescu and Y. Chisti, *Biotechnol. Adv.*, 2005, **23**, 471–499.
26. L. Johnsson, M. Hokeberg and B. Gerhardson, *Eur. J. Plant Pathol.*, 1998, **104**, 701–711.
27. M. Gavrilescu and Y. Chisti, *Biotechnol. Adv.*, 2005, **23**, 471–499.
28. G. J. Ash, *Biol. Control*, 2010, **52**, 230–240.
29. G. Q. Chen and R. Kazlauskas, *Curr. Opin. Biotechnol.*, 2011, **22**, 747–748.
30. K. E. Jaeger and P. Holliger, *Curr. Opin. Biotechnol.*, 2010, **21**, 711–712.
31. K. Watanabe and G. Bennett, *Curr. Opin. Biotechnol.*, 2009, **20**, 607–609.
32. J. Kuc, *Eur. J. Plant Pathol.*, 2001, **107**, 7–12.
33. J. L. Dangl, D. M. Horvath and B. J. Staskawicz, *Science*, 2013, **341**, 746–751.
34. J. D. Jones and J. L. Dangl, *Nature*, 2006, **444**, 323–329.
35. U. Conrath, O. Thulke, V. Katz, S. Schwindling and A. Kohler, *Eur. J. Plant Pathol.*, 2001, **107**, 113–119.
36. K. Goellner and U. Conrath, *Eur. J. Plant Pathol.*, 2008, **121**, 233–242.
37. C. M. Pieterse, A. Leon-Reyes, S. Van der Ent and S. C. Van Wees, *Nat. Chem. Biol.*, 2009, **5**, 308–316.

38. D. Haas and G. Defago, *Nat. Rev. Microbiol.*, 2005, **3**, 307–319.
39. J. D. G. Jones and J. L. Dangl, *Nature*, 2006, **444**, 323–329.
40. G. J. Beckers and U. Conrath, *Curr. Opin. Plant Biol.*, 2007, **10**, 425–431.
41. R. F. White, *Virology*, 1979, **99**, 410–412.
42. M. M. Lopez, M. J. Lopez-Lopez, R. Marti, J. Zamora, J. Lopez-Sanchez and R. Beltra, *Potato Res.*, 2001, **44**, 197–206.
43. M. Rivas-San Vicente and J. Plasencia, *J. Exp. Bot.*, 2011, **62**, 3321–3338.
44. S. Pasqualini, G. Della Torre, F. Ferranti, L. Ederli, C. Piccioni, L. Reale and M. Antonielli, *Physiol. Plant.*, 2002, **115**, 204–212.
45. J. Siegrist, M. Orober and H. Buchenauer, *Physiol. Mol. Plant Pathol.*, 2000, **56**, 95–106.
46. N. Yalpani, D. J. Altier, E. Barbour, A. L. Cigan and C. J. Scelonge, *Plant Cell*, 2001, **13**, 1401–1409.
47. A. Molina, S. Volrath, D. Guyer, K. Maleck, J. Ryals and E. Ward, *Plant J.*, 1999, **17**, 667–678.
48. D. Sandhu, I. M. Tasma, R. Frasch and M. K. Bhattacharyya, *BMC Plant Biol.*, 2009, **9**, 105, DOI: 10.1186/1471-2229-9-105.
49. S. P. Yan, W. Wang, J. Marques, R. Mohan, A. Saleh, W. E. Durrant, J. Q. Song and X. N. Dong, *Mol. Cell*, 2013, **52**, 602–610.
50. H. Suzuki, M. S. S. Reddy, M. Naoumkina, N. Aziz, G. D. May, D. V. Huhman, L. W. Sumner, J. W. Blount, P. Mendes and R. A. Dixon, *Planta*, 2005, **220**, 696–707.
51. S. Di Marco, F. Osti, F. Calzarano, R. Roberti, A. Veronesi and C. Amalfitano, *Phytopathol. Mediterr.*, 2011, **50**, S285–S299.
52. J. J. Noh, W. Kim, K. K. Lee, S. Y. So, B. R. Ko and D. H. Kim, *Korean J. Hortic. Sci. Technol.*, 2007, **25**, 24–28.
53. C. F. Zhang, S. X. Li, B. Q. Li and Q. Zhang, *The Effects of Humic Substances on the Activity of Metalaxyl Plus Mancozeb, Fungicide*, 1999.
54. A. O. Jackson and C. B. Taylor, *Plant Cell*, 1996, **8**, 1651–1668.
55. J.-g. Liang, R.-x. Tao, Z.-n. Hao, L.-p. Wang and X. Zhang, *Afr. J. Biotechnol.*, 2011, **10**, 6920–6927.
56. T.-C. Lin, M. Ishizaka and H. Ishii, *J. Phytopathol.*, 2009, **157**, 40–50.
57. Y. Narusaka, M. Narusaka, T. Horio and H. Ishii, *Ann. Phytopathol. Soc. Jpn.*, 1999, **65**, 116–122.
58. S. Bautista-Banos, A. N. Hernandez-Lauzardo, M. G. Velazquez-del Valle, M. Hernandez-Lopez, E. A. Barka, E. Bosquez-Molina and C. L. Wilson, *Crop Prot.*, 2006, **25**, 108–118.
59. K. F. Zeng, Y. Y. Deng, J. A. Ming and L. L. Deng, *Sci. Hortic.*, 2010, **126**, 223–228.
60. I. S. Sobhy, M. Erb, A. A. Sarhan, M. M. El-Husseini, N. S. Mandour and T. C. J. Turlings, *J. Chem. Ecol.*, 2012, **38**, 348–360.
61. N. I. Vasyukova, G. I. Chalenko, T. A. Valueva, N. G. Gerasimova, Y. S. Panina and O. L. Ozeretskovskaya, *Appl. Biochem. Microbiol.*, 2003, **39**, 613–617.
62. M. Krause and J. Durner, *Mol. Plant-Microbe Interact.*, 2004, **17**, 131–139.
63. T. S. Workneh, G. Osthoff and M. Steyn, *J. Food Sci. Technol.*, 2012, **49**, 685–694.

64. S. C. M. van Wees and J. Glazebrook, *Plant J.*, 2003, **33**, 733–742.
65. K. Ament, V. Krasikov, S. Allmann, M. Rep, F. L. Takken and R. C. Schuurink, *Plant J.*, 2010, **62**, 124–134.
66. S. W. Park, P. P. Liu, F. Forouhar, A. C. Vlot, L. Tong, K. Tietjen and D. F. Klessig, *J. Biol. Chem.*, 2009, **284**, 7307–7317.
67. Y. Choh, R. Ozawa and J. Takabayashi, *Appl. Entomol. Zool.*, 2004, **39**, 311–314.
68. T. Shoji, T. Ogawa and T. Hashimoto, *Plant Cell Physiol.*, 2008, **49**, 1003–1012.
69. J. Zhao, S. H. Zheng, K. Fujita and K. Sakai, *J. Exp. Bot.*, 2004, **55**, 1003–1012.
70. A. Y. Zhang, M. Y. Jiang, J. H. Zhang, M. P. Tan and X. L. Hu, *Plant Physiol.*, 2006, **141**, 475–487.
71. Y. Araki and Y. Kurahashi, *J. Pestic. Sci.*, 1999, **24**, 369–374.
72. P. G. Justyna and K. Ewa, *Acta Physiol. Plant.*, 2013, **35**, 1735–1748.
73. Z. K. Zhang, D. P. Yang, B. Yang, Z. Y. Gao, M. Li, Y. M. Jiang and M. J. Hu, *Sci. Hortic.*, 2013, **160**, 78–84.
74. L. Zimmerli, C. Jakab, J. P. Metraux and B. Mauch-Mani, *Proc. Natl. Acad. Sci. U. S. A.*, 2000, **97**, 12920–12925.
75. Y. M. Li, Z. K. Zhang, Y. T. Jia, Y. M. Shen, H. M. He, R. X. Fang, X. Y. Chen and X. J. Hao, *Plant Biotechnol. J.*, 2008, **6**, 301–308.
76. M. F. Abdel-Monaim, *Afr. J. Biotechnol.*, 2011, **10**, 10842–10855.
77. M. Nishioka, H. Nakashita, H. Suzuki, S. Akiyama, S. Yoshida and I. Yamaguchi, *J. Pestic. Sci.*, 2003, **28**, 416–421.
78. M. Yasuda, M. Nishioka, H. Nakashita, I. Yamaguchi and S. Yoshida, *Biosci., Biotechnol., Biochem.*, 2003, **67**, 2614–2620.
79. H. Turkusay, N. Tosun, S. Yildiz and H. Saygili, in *II International Symposium on Tomato Diseases*, ed. H. Saygili, F. Sahin and Y. Aysan, 2009, vol. 808, pp. 431–435.
80. T. Iwai, S. Seo, I. Mitsuhara and Y. Ohashi, *Plant Cell Physiol.*, 2007, **48**, 915–924.
81. H. Kauss, E. Theisinger-Hinkel, R. Mindermann and U. Conrath, *Plant J.*, 1992, **2**, 655–660.
82. T. Maeda and H. Ishiwari, *Exp. Appl. Acarol.*, 2012, **58**, 247–258.
83. Y. F. Xu, Z. J. Zhao, X. H. Qian, Z. G. Qian, W. H. Tian and J. J. Zhong, *J. Agric. Food Chem.*, 2006, **54**, 8793–8798.
84. Q. S. Du, W. P. Zhu, Z. J. Zhao, X. H. Qian and Y. F. Xu, *J. Agric. Food Chem.*, 2012, **60**, 346–353.
85. G. E. Vallad and R. M. Goodman, *Crop Sci.*, 2004, **44**, 1920–1934.
86. W. Kunz, R. Schurter and T. Maetzke, *Pestic. Sci.*, 1997, **50**, 275–282.
87. Q. S. Du, Y. X. Shi, P. F. Li, Z. J. Zhao, W. P. Zhu, X. H. Qian, B. J. Li and Y. F. Xu, *Chin. Chem. Lett.*, 2013, **24**, 967–969.
88. V. Flors, C. Miralles, C. Gonzalez-Bosch, M. Carda and P. Garcia-Agustin, *Physiol. Mol. Plant Pathol.*, 2003, **63**, 151–158.
89. F. L. Liu, Z. J. Fan, H. B. Song, X. F. Liu and Y. G. Zhang, *Acta Crystallogr., Sect. E: Struct. Rep. Online*, 2005, **61**, O4054–O4055.

90. H. Nakashita, M. Yasuda, M. Nishioka, S. Hasegawa, Y. Arai, M. Uramoto, S. Yoshida and I. Yamaguchi, *Plant Cell Physiol.*, 2002, **43**, 823–831.

91. Q. Du, W. Zhu, Z. Zhao, X. Qian and Y. Xu, *J. Agric. Food Chem.*, 2011, **60**, 346–353.

92. P. Stanetty, M. Kremslehner and M. Jaksits, *Pestic. Sci.*, 1998, **54**, 316–319.

93. P. Stanetty, M. Kremslehner and H. Vollenkle, *J. Chem. Soc., Perkin Trans. 1*, 1998, 853–856.

94. Q.-S. Du, Y.-X. Shi, P.-F. Li, Z.-J. Zhao, W.-P. Zhu, X.-H. Qian, B.-J. Li and Y.-F. Xu, *Chin. Chem. Lett.*, 2013, **24**, 967–969.

95. G. E. Vallad and R. M. Goodman, *Crop Sci.*, 2004, **44**, 1920–1934.

96. S. L. Willingham, K. G. Pegg, P. W. B. Langdon, A. W. Cooke, D. Peasley and R. McLennan, *Australas. Plant Pathol.*, 2002, **31**, 333–336.

97. F. Pasquer, E. Isidore, J. Zarn and B. Keller, *Plant Mol. Biol.*, 2005, **57**, 693–707.

98. T. H. D. Thi, R. C. Puig, H. K. Kim, C. Erkelens, A. W. M. Lefeber, H. J. M. Linthorst, Y. H. Choi and R. Verpoorte, *Plant Physiol. Biochem.*, 2009, **47**, 146–152.

99. M. van Hulten, M. Pelser, L. C. van Loon, C. M. J. Pieterse and J. Ton, *Proc. Natl. Acad. Sci. U. S. A.*, 2006, **103**, 5602–5607.

100. S. Ziadi, S. Barbedette, J. F. Godard, C. Monot, D. Le Corre and D. Silue, *Plant Pathol.*, 2001, **50**, 579–586.

101. J. Görlach, S. Volrath, G. Knauf-Beiter, G. Hengy, U. Beckhove, K. H. Kogel, M. Oostendorp, T. Staub, E. Ward, H. Kessmann and J. Ryals, *Plant Cell*, 1996, **8**, 629–643.

102. H. X. Liu, W. B. Jiang, Y. Bi and Y. B. Luo, *Postharvest Biol. Biotechnol.*, 2005, **35**, 263–269.

103. J. Cao, W. Jiang and H. He, *J. Phytopathol.*, 2005, **153**, 640–646.

104. A. Buzi, G. Chilosi, D. De Sillo and P. Magro, *J. Phytopathol.*, 2004, **152**, 34–42.

105. Australian Pesticides and Veterinary Medicines Authority, APVMA, *Evaluation of the new active acibenzolar-S-methyl in the product Bion plant activator seed treatment*, Canberra, Australia, 2007.

106. A. I. Bokshi, S. C. Morris, R. M. McConchie and B. J. Deverall, *J. Hortic. Sci. Biotechnol.*, 2006, **81**, 700–706.

107. U. Conrath, C. M. J. Pieterse and B. Mauch-Mani, *Trends Plant Sci.*, 2002, **7**, 210–216.

108. V. A. Halim, L. Eschen-Lippold, S. Altmann, M. Birschwilks, D. Scheel and S. Rosahl, *Mol. Plant-Microbe Interact.*, 2007, **20**, 1346–1352.

109. A. Anand, S. R. Uppalapati, C. M. Ryu, S. N. Allen, L. Kang, Y. H. Tang and K. S. Mysore, *Plant Physiol.*, 2008, **146**, 703–715.

110. Z. G. Qian, Z. J. Zhao, Y. F. Xu, X. H. Qian and J. J. Zhong, *Appl. Microbiol. Biotechnol.*, 2006, **71**, 164–167.

111. K. Umemura, J. Satou, M. Iwata, N. Uozumi, J. Koga, T. Kawano, T. Koshiba, H. Anzai and M. Mitomi, *Plant J.*, 2009, **57**, 463–472.

112. M. Yasuda, H. Nakashita, S. Hasegawa, M. Nishioka, Y. Arai, M. Uramoto, I. Yamaguchi and S. Yoshida, *Biosci., Biotechnol., Biochem.*, 2003, **67**, 322–328.

113. H. Nakashita, K. Yoshioka, M. Takayama, R. Kuga, N. Midoh, R. Usami, K. Horikoshi, K. Yoneyama and I. Yamaguchi, *Biosci., Biotechnol., Biochem.*, 2001, **65**, 205–208.

114. J. Yu, J. O. Gao, X. Y. Wang, Q. A. Wei, L. F. Yang, K. Qiu and B. K. Kuai, *J. Plant Biol.*, 2010, **53**, 417–424.

115. T. Mahmood, M. Kakishima and S. Komatsu, *Protein Pept. Lett.*, 2009, **16**, 1041–1052.

116. X. Yi and Y. Lu, *Chemosphere*, 2006, **65**, 639–643.

117. M. Iwata, *Pestic. Outlook*, 2001, **12**, 28–31.

118. M. Katohgi, H. Togo, K. Yamaguchi and M. Yokoyama, *Tetrahedron*, 1999, **55**, 14885–14900.

119. C. Boyle and D. R. Walters, *Plant Pathol.*, 2006, **55**, 82–91.

120. M. Sugimori, K. Kiribuchi, C. Akimoto, T. Yamaguchi, E. Minami, N. Shibuya, H. Sobajima, E. M. Cho, N. Kobashi, H. Nojiri, T. Omori, M. Nishiyama and H. Yamane, *Biosci., Biotechnol., Biochem.*, 2002, **66**, 1140–1142.

121. H. Sobajima, T. Tani, T. Chujo, K. Okada, K. Suzuki, S. Mori, E. Minami, M. Nishiyama, H. Nojiri and H. Yamane, *Biosci., Biotechnol., Biochem.*, 2007, **71**, 3110–3115.

122. N. Tanaka, F. S. Che, N. Watanabe, S. Fujiwara, S. Takayama and A. Isogai, *Mol. Plant-Microbe Interact.*, 2003, **16**, 422–428.

123. H. Nakashita, K. Yoshioka, M. Yasuda, T. Nitta, Y. Arai, S. Yoshida and I. Yamaguchi, *Physiol. Mol. Plant Pathol.*, 2002, **61**, 197–203.

124. K. Yoshioka, H. Nakashita, D. F. Klessig and I. Yamaguchi, *Plant J.*, 2001, **25**, 149–157.

125. S. Komatsu, G. Yang, N. Hayashi, H. Kaku, K. Umemura and Y. Iwasaki, *Plant, Cell Environ.*, 2004, **27**, 947–957.

126. Y. Y. Lu, Y. H. Liu and C. Y. Chen, *Plant Sci.*, 2007, **172**, 913–919.

127. K. Sakamoto, Y. Tada, Y. Yokozeki, H. Akagi, N. Hayashi, T. Fujimura and N. Ichikawa, *Plant Mol. Biol.*, 1999, **40**, 847–855.

128. K. Tsubata, K. Kuroda, Y. Yamamoto and N. Yasokawa, *J. Pestic. Sci.*, 2006, **31**, 161–162.

129. M. Yasuda, H. Nakashita and S. Yoshida, *J. Pestic. Sci.*, 2004, **29**, 46–49.

130. M. Yasuda, M. Kusajima, M. Nakajima, K. Akutsu, T. Kudo, S. Yoshida and H. Nakashita, *J. Pestic. Sci.*, 2006, **31**, 329–334.

131. M. Yasuda, *J. Pestic. Sci.*, 2007, **32**, 281–282.

132. Q. X. Zheng, N. Mi, Z. J. Fan, X. A. Zu, H. K. Zhang, H. A. Wang and Z. K. Yang, *J. Agric. Food Chem.*, 2010, **58**, 7846–7855.

133. N. Chidananda, B. Poojary, V. Sumangala, N. S. Kumari, P. Shetty and T. Arulmoli, *Eur. J. Med. Chem.*, 2012, **51**, 124–136.

134. N. B. Sun, J. Q. Fu, J. Q. Weng, J. Z. Jin, C. X. Tan and X. H. Liu, *Molecules*, 2013, **18**, 12725–12739.

135. H. Kaur and S. Kumar, *Asian J. Chem.*, 2000, **12**, 629–632.

136. S. X. Wang, X. Zuo, Z. J. Fan, Z. C. Zhang, J. F. Zhang, L. X. Xiong, Y. F. Fu, Z. Fang, Q. J. Wu and Y. J. Zhang, *Chin. J. Org. Chem.*, 2013, **33**, 2367–2375.

137. Q. Zheng, N. Mi, Z. Fan, X. Zuo, H. Zhang, H. Wang and Z. Yang, *J. Agric. Food Chem.*, 2010, **58**, 7846–7855.

138. H. Wang, Z. Yang, Z. Fan, Q. Wu, Y. Zhang, N. Mi, S. Wang, Z. Zhang, H. Song and F. Liu, *J. Agric. Food Chem.*, 2011, **59**, 628–634.

139. Z. Fan, Z. Shi, H. Zhang, X. Liu, L. Bao, L. Ma, X. Zuo, Q. Zheng and N. Mi, *J. Agric. Food Chem.*, 2009, **57**, 4279–4286.

140. K. Wang, B. Su, Z. Wang, M. Wu, Z. Li, Y. Hu, Z. Fan, N. Mi and Q. Wang, *J. Agric. Food Chem.*, 2010, **58**, 2703–2709.

141. X. Zuo, N. Mi, Z. Fan, Q. Zheng, H. Zhang, H. Wang and Z. Yang, *J. Agric. Food Chem.*, 2010, **58**, 2755–2762.

142. Z. Fan, Z. Yang, H. Zhang, N. Mi, H. Wang, F. Cai, X. Zuo, Q. Zheng and H. Song, *J. Agric. Food Chem.*, 2010, **58**, 2630–2636.

143. Z. Fan, *J. Pestic. Sci.*, 2011, **36**, 162.

144. M. Dicke, *Entomol. Exp. Appl.*, 1999, **91**, 131–142.

145. C. A. M. Robert, M. Erb, B. E. Hibbard, B. W. French, C. Zwahlen and T. C. J. Turlings, *Funct. Ecol.*, 2012, **26**, 1429–1440.

146. H. H. Cao, S. H. Wang and T. X. Liu, *Insect Sci.*, 2014, **21**, 47–55.

147. S. H. Chung, C. Rosa, E. D. Scully, M. Peiffer, J. F. Tooker, K. Hoover, D. S. Luthe and G. W. Felton, *Proc. Natl. Acad. Sci. U. S. A.*, 2013, **110**, 15728–15733.

148. J. Kastner, D. von Knorre, H. Himanshu, M. Erb, I. T. Baldwin and S. Meldau, *PLoS One*, 2014, **9**.

149. I. A. Vos, A. Verhage, R. C. Schuurink, L. G. Watt, C. M. J. Pieterse and S. C. M. Van Wees, *Front. Plant Sci.*, 2013, **4**, 539.

150. W. C. Cooper, L. Jia and F. L. Goggin, *J. Chem. Ecol.*, 2004, **30**, 2527–2542.

151. G. Nombela, S. Pascual, M. Aviles, E. Guillard and M. Muniz, *J. Econ. Entomol.*, 2005, **98**, 2266–2271.

152. T. Maeda and H. Ishiwari, *Exp. Appl. Acarol.*, 2012, **58**, 247–258.

153. C. L. Truitt and P. W. Pare, *Planta*, 2004, **218**, 999–1007.

154. M. Rostas and T. C. J. Turlings, *Biol. Control*, 2008, **46**, 178–186.

155. Y. Noutoshi, M. Okazaki, T. Kida, Y. Nishina, Y. Morishita, T. Ogawa, H. Suzuki, D. Shibata, Y. Jikumaru, A. Hanada, Y. Kamiya and K. Shirasu, *Plant Cell*, 2012, **24**, 3795–3804.

156. E. K. Seo, H. Nakamura, M. Mori and T. Asami, *Bioorg. Med. Chem. Lett.*, 2012, **22**, 1761–1765.

157. W. Lu, X. Liu, X. Cao, M. Xue, K. Liu, Z. Zhao, X. Shen, H. Jiang, Y. Xu, J. Huang and H. Li, *J. Med. Chem.*, 2011, **54**, 3564–3574.

Subject Index